Pest Management in Museums, Archives and Historic Houses

Pest Management in Museums, Archives and Historic Houses

David Pinniger

Illustrated by Annette Townsend
Rodent and bird chapter by Adrian Meyer

Archetype Publications

First published 2001 by Archetype Publications Ltd.

Archetype Publications Ltd.
6 Fitzroy Square
London W1T 5HJ

www.archetype.co.uk

Tel: 44(207) 380 0800
Fax: 44(207) 380 0500

ISBN 1-873132-86-7

Typeset by Kate Williams, Abergavenny
Printed and bound in Great Britain by Henry Ling Ltd, The Dorset Press, Dorchester

Contents

Acknowledgements

I would like to thank all my friends and colleagues in museums, universities and historic houses in many towns and countries. Much of the information in this book has been based on practical experience gained with their help and support. I would particularly like to thank Bob Child for sharing his knowledge and bad jokes and Adrian Meyer for enabling the book to include rodents and birds. I would also pay tribute to the skill, patience and tolerance of Annette Townsend whose beautiful drawings enhance this edition. Finally, thanks to Geoff Stansfield who started me off with the idea.

This book is for Edward Pinniger, Edwin Bezant and Alec Green who all played their part in making me an entomologist.

Preface

This book has its origin in course notes written for Summer School courses run by International Academic Projects at the Institute of Archaeology in London. These were compiled into the first edition of *Insect Pests in Museums* (Institute of Archaeology Publications 1989). The third, much improved and revised edition was published by Archetype Publications in 1994.

Since then, there have been many developments in what has become known as Integrated Pest Management (IPM) in the heritage sector. A major influence for change has been the concern regarding the use of pesticides and their effects on objects, staff and environment. It has been recognised that IPM applied to museums, archives and historic houses is not only safer but also more cost-effective. This new book incorporates much of the material contained in *Insect Pests in Museums* but it has been expanded to include as much up-to-date information as possible. It is intended to be a working guide to help people recognise pests and take practical steps to prevent and control damage to collections. It includes new information on trapping and detection of pests, and evaluates the advantages and disadvantages of physical and chemical control measures. The illustrations by Annette Townsend were commissioned especially for this book and I believe they are some of the clearest pictures of museum insects yet published. There is also a brand new chapter on rodent and bird pests written by Adrian Meyer, an acknowledged world expert in the field. I hope that readers will find the book useful in the prevention and control of pest problems. I am always interested to receive any information from readers that will improve the coverage of pests and IPM methods for any future edition.

David Pinniger
May 2001

CHAPTER ONE

Integrated Pest Management in museums

Introduction

Pests come into conflict with man because they compete with us for resources. Ever since man first started to store food or make clothing and other artefacts he has had a battle. Increasing affluence and the creation of religious and artistic objects added another area of conflict with pests. In addition to destruction of property, some pests, such as rats and mice, have carried diseases that have wiped out millions of people.

All too often, pest control in museums and historic houses is a reaction to the discovery of insect activity and damage. The aim of this book is to provide practical, safe and cost-effective advice on the prevention and control of pests. It will explain what a museum needs to do to protect collections, furnishings and buildings from harm by pests, and it describes the options for controlling infestations and strategies to reduce the risks to collections.

What is IPM ?

Integrated Pest Management (IPM) is a term first adopted to describe the development of new pest control methods for fruit and crops that did not rely on continuous use of pesticides. We have successfully adapted the main IPM principles for use in museums and houses: monitoring for pests, targeting treatment only where it is needed and modifying the environment to discourage pest attack.

Why IPM in a museum?

Care of collections and historic buildings involves many different disciplines including conservation and management of collections and buildings. The major influences on deterioration are the environmental effects of light, humidity and agents of decay such as insects and moulds. All of these factors are interrelated and IPM should seek to approach all pest problems using a holistic approach rather than simply as a reaction to each individual crisis (Butcher-Youghans and Anderson 1990; Stansfield 1985). The expertise in collections care in the museum world can be applied to the development of an IPM programme tailored to the specific needs of a collection or historic house. A well-planned and executed IPM programme will prevent the occurrence of problems and crises (Rossol and Jessup 1996), as well as making more effective use of limited human and cash resources.

Developing an IPM strategy

IPM should be relevant to the needs of the museum, building or collection and should use as much local information and expertise as possible. It should also be practical and achievable – it is all too easy to devise a grandiose IPM scheme that then turns out to be unworkable. It should be a process of evolution rather than revolution, and encourage participation by different disciplines.

The key stages are:

1. Recognising and identifying priorities for action.
2. Identifying responsible staff.
3. Taking action on the high priorities.
4. Establishing procedures for forward planning, financing and review.

In order to develop an IPM strategy it is important to understand and recognise some of the key components of successful pest control:

- Avoid pests – by keeping them out.
- Prevent pests – by denying them safe haven.
- Recognise pests – the main species and the damage they cause.
- Assess the problem – by inspection and trapping.
- Solve pest problems – by improving the environment and carrying out appropriate treatments.
- Review IPM procedures periodically and change them when necessary to improve the strategy.

Preventing pests

Most insect pest problems in collections are caused by beetles and moths although other insects such as termites, cluster flies, ants and silverfish can cause damage and nuisance. Rodents will gnaw objects and mice can be very damaging once they start to breed within a building. Bird droppings, particularly from pigeons, can cause fouling and deterioration and birds' nests will harbour many species of insect pests. Dead rodents and birds will in turn provide food for insects. A key to avoiding pests is understanding what makes them thrive so we can prevent them becoming established and causing damage.

Many modern buildings have been successfully proofed to prevent the entry of pests but keeping insects, birds and rodents out of older and historic buildings is often far more difficult because of the unacceptable appearance of some proofing methods. Simple methods – such as sealing of windows, bristle strips on doors and wire mesh in eaves and downpipes – can reduce dramatically the risk of pests infiltrating indoors.

It is not usually possible to exclude pests completely from a building and so it is important that they are denied a suitable environment if they do get in. The four key things needed by pests are: food; harbourage; warmth; water or humidity.

Food and harbourage

Because insects are small they may find sufficient food in relatively small areas that may not be immediately obvious. Because of this, cleaning is the most important part of any IPM programme. Unused rooms and storage areas are often neglected and dirt and debris will provide an ideal harbourage for insects. The unavailability of human food is usually the most important limiting factor for bird and rodent pests so good hygiene and management of catering is crucial to preventing problems.

Warmth

Warm temperatures of 20°C and above will encourage insect breeding and so cool conditions are recommended. Although it may not be possible to lower temperatures in public areas, storage should be operated at as low a temperature as possible. Direct sunlight can cause localised hot-spots even in cool areas and uneven temperatures may result in localised condensation.

Water or humidity

Birds and rodents usually need access to drinking water and so leaks from gutters or waterpipes should be checked and remedied. Some insects will survive at low humidities but many species thrive when it is damper. Furniture beetle or woodworm will only successfully complete its life cycle when wood is in an environment above 60% relative humidity (RH). Silverfish will only breed rapidly and cause serious problems in conditions of above 70% RH. Booklice also need higher levels of RH than is normally found in libraries and archives and so they are often found in damp basements or in localised damp areas. RH levels should be measured and monitored, and sources of dampness such as condensation or poor damp-proofing should be identified and rectified.

Recognising pests

Rodent pests

There are only two species in the UK that cause serious problems, easily identified by their profuse droppings, gnaw marks and habits. Serious infestations of brown rats, *Rattus norvegicus*, are rare in museums but may be more common in outbuildings in rural areas. In buildings they are usually associated with considerable and accessible sources of human food, poor refuse disposal systems and poor sanitation. The house mouse, *Mus domesticus*, is the most common pest, although in many museums and houses, the fear of infestation is often greater than the reality. If they have undisturbed areas with access to food supplies, then mice will breed rapidly and create serious damage by shredding paper and textiles to make nest bedding – they will not discriminate between valuable objects, packing and rubbish.

Bird pests

Any nesting or roosting bird can be a potential source of problems. Feral pigeons are the most common pest in urban areas when they roost or nest in buildings. Starlings, sparrows, gulls and rooks may also cause serious problems of roosting and fouling in specific locations.

Insects

There are many species of insects with wide-ranging habits and food preferences (see Table 1 for a list of insect pests found in museums). They will attack and destroy wood, paper, textiles, natural history specimens and many ethnographic materials. Although it is often the adult stage of an insect that is found, it is the larval stage that causes the most damage. Adults will usually be more active and obvious in the summer months, but larvae hatched from eggs laid by the adults will feed and grow throughout the rest of the year.

Many insects found in museums are not pests but have simply strayed into the building. Although the dead bodies of flies, wasps and ground beetles may not present a direct risk to the collection, they can provide a substantial source of food for real pests such as carpet beetles.

Assessing the problem

Before taking action it is useful to draw up a check list to determine the priorities:

1. Are there signs of insects, rodents or birds?
2. What species?
3. Are they alive or dead?
4. How many are there?
5. Are they casual visitors or breeding in the building?
6. Has damage been caused to objects?
7. How many objects are affected?
8. Are they in display material?
9. Is there damage to the building?
10. Are they elsewhere in the building?

A thorough survey of the building is essential after the initial assessment. Look for any signs of infestation and inspect as many places as possible. Unused rooms and dead spaces are particularly favoured by pests. Look for insects and rodent droppings in dark areas using a good flashlight. Look for signs of insect activity such as fresh frass (debris of food fragments and dry excreta) from woodworm exit holes, moth webbing or cast skins of carpet beetle woolly bears. Examine windowsills and light fittings in spring and summer for signs of adult insects.

Education and school liaison sections can present pest problems because they may use de-accessioned objects and infestable materials such as pasta and wool felt for creative and hands-on sessions with school groups.

Remember to check all food storage and preparation areas including canteens, cafeterias and mess rooms as these are a major focus for rodent pests.

Table 1 Insect pests found in museums: common and scientific (Latin) names

Scientific name	Common name
Beetles	
Anobium punctatum	Common furniture beetle (woodworm)
Anthrenus flavipes	Furniture carpet beetle
Anthrenus sarnicus	Guernsey carpet beetle
Anthrenus verbasci	Varied carpet beetle
Attagenus pellio	Two-spot carpet beetle (fur beetle)
Attagenus smirnovi	Brown carpet beetle (vodka beetle)
Attagenus unicolor (*megatoma*)	Black carpet beetle
Dermestes maculatus	Hide beetle
Dermestes peruvianus	Peruvian hide beetle
Dinoderus sp.	Bamboo borer
Euophryum confine	Woodboring weevil
Gibbium psylloides	Shiny spider beetle
Hylotrupes bajulus	House longhorn beetle
Lasioderma serricorne	Cigarette beetle
Lyctus brunneus	Powder post beetle
Nacerdes melanura	Wharf borer
Niptus hololeucus	Golden spider beetle
Ptinus fur	White-marked spider beetle
Ptinus tectus	Australian spider beetle
Reesa vespulae	Museum nuisance/American wasp beetle
Scolytus sp	Bark beetle
Sinoxylon anale	Auger borer
Stegobium paniceum	Biscuit beetle or drugstore beetle
Tenebrio molitor	Mealworm beetle
Thylodrias contractus	Odd beetle
Trogoderma angustum	Cabinet beetle
Xestobium rufovillosum	Death watch beetle
Moths	
Endrosis sarcitrella	White-shouldered house moth
Hofmannophila pseudospretella	Brown house moth
Tinea pellionella	Case-bearing clothes moth
Tineola bisselliella	Webbing clothes moth
Trichophaga tapetzella	Tapestry moth
Other insects	
Blatta orientalis	Oriental cockroach
Blattella germanica	German cockroach
Periplaneta americana	American cockroach
Liposcelis bostrychophila	Booklouse or psocid
Lepisma saccharina	Silverfish or fishmoth
Thermobia domestica	Firebrat
Lasius niger	Black garden ant
Monomorium pharaonis	Pharaoh's ant
Camponotus herculeanus	Hercules carpenter ant
Pollenia rudis	Cluster fly
Reticulitermes sp.	Subterranean termite
Macrotermes sp.	Subterranean termite
Cryptotermes sp.	Drywood termite
Kalotermes sp.	Drywood termite
Mites	
Acarus siro	Flour mite
Bryobia praetiosa	Red clover mite
Glycyphagus domesticus	House mite

Surveying collections is usually far more difficult because it is normally impossible to inspect all objects, particularly when they are in storage. It is therefore essential to make a priority list of objects containing materials which are far more vulnerable than others.

Materials most at risk from insect attack

- Fur
- Hair
- Feathers
- Wool
- Silk
- Insect specimens
- Parchment and vellum
- Papier mâché
- Animal skins
- Dried food such as pasta or flour
- Sapwoods (above 60% RH)
- Dried plants and seeds
- Any damp organic material

Some materials such as cotton and synthetic fabrics are not normally attacked, and paper and books are rarely attacked unless they are dirty and damp. In general, dirty and neglected objects in dark places will be more at risk than those that are clean and positioned in well-lit areas.

Quarantine

An essential part of any pest prevention policy in a museum is to keep pests out of collections. Insects can be introduced from many sources including new acquisitions, objects on loan from other museums and objects returned from loan.

Objects must be checked for infestation before being allowed into the main collection area, whether in store or on display. Inspection may reveal insect damage and moth webbing but insect eggs or small larvae may be difficult to see. Woodborer emergence holes may be obvious but developing larvae will be hidden in the wood. This means that some incubation period may be necessary to determine whether an infestation is active or long dead.

A quarantine area must be provided and all staff involved with handling and movement of objects must be aware of the quarantine strategy. Any breach can result in infestation being introduced undetected into collections.

Solving pest problems

Prevention is always better than cure, but if pests are found in objects or in the building then some remedial action may be necessary. The options for the most appropriate course of action vary depending upon the pest and whether it is in the building or in objects (see Chapter 5).

1. Isolate any objects suspected of being infested with insects to prevent spread of infestation to other objects.
2. Clean infested areas and destroy insect bodies and debris.
3. Clean up and destroy all rodent droppings and gnawed material.
4. Decide on the most appropriate treatment for the object and environment.

Pests in the building environment

For rodents, target the infestation using traps or rodenticides (see Chapter 6). For insects, use an approved insecticidal dust or spray (see Chapter 3). Insects which live in harbourages will only be killed when they wander across treated surfaces.

Rodent traps and localised insecticide treatments can be safely used or carried out by museum staff but rodenticide treatments and large-scale insecticide treatments should be carried out by a competent contractor.

Insects in objects

The choice of remedial treatment will depend upon the severity of the infestation, the type of material and the value of the object. Treatment of objects should only be carried out after checking with a conservator or collections care specialist.

Fumigation with a toxic gas such as methyl bromide or phosphine was formerly the principal method of killing insects in objects. Methyl bromide is being withdrawn from use because it is an ozone-depleter and residues or reactions from other fumigant gases are now much less acceptable. Recent developments mean that museums now have a number of alternatives to toxic gases which will kill all pests in objects if they are carried out correctly. These include low temperature, elevated temperature, carbon dioxide and nitrogen anoxia (see Chapter 3). Both freezing and nitrogen treatments can be safely carried out by museum staff with appropriate training and experience.

Implementing IPM

An IPM programme is essentially an evolving process which should apply local intelligence and respond to changing needs and priorities. Implementing an IPM programme in a large museum or collection can be a daunting task. It is therefore important to identify priorities and plan to cover the museum in achievable steps. In many cases it has taken some years to develop and implement a programme in a large national collection. It has been achieved by initially selecting one department or collection area, demonstrating that IPM is effective and achievable in that area and then adapting the programme for other areas until all parts are covered.

Successful IPM involves all staff. Training must be appropriate and comprehensive to enable staff to identify problems and make decisions on appropriate action. One person should be identified as the co-ordinator and chief contact for IPM matters.

The problem of budgeting for IPM should not be underestimated as funding becomes increasingly difficult. It is essential to find out what is currently being spent on existing pest control contracts and any fumigation treatments. In many cases, a good IPM programme can use this money more effectively and efficiently. It is also important to consider the cost of collecting and conservation – IPM will avoid neglect and deterioration of objects and the consequence of increased expenditure on conservation.

Because of the effects of some chemicals on staff, objects and the environment there will be increased pressure to move away from persistent and toxic insecticides. The development of an IPM programme based on the principles outlined in this book will enable collections and buildings to be cared for in ways that are safe and effective. See Chapter 7 for further information on implementing IPM.

CHAPTER TWO

Insect pests

Why are insects pests?

Most insects are not pests: of the hundreds of thousands of different species in the world, only a very small percentage come into direct conflict with man. The competition is between museums and houses that are trying to conserve and maintain their collections, and the insects that will damage, destroy or despoil objects and buildings. The most serious insect pests were not always pests, and we need to look at natural habitats to see where they originated.

Most of the textile pests, such as clothes moths and carpet beetles, occupy a natural environmental niche in bird and mammal nests. Here they perform a very important role in the breakdown and recycling of hair, skin, feathers and excreta. Pests of stored food, such as grain weevils and biscuit beetles, can be found in the caches of food which some animals store over the autumn and winter. Some pests have been associated with man for many years. For example, clothes moths caused problems for the Ancient Greeks and Romans and are mentioned in the Bible, woodborers were probably stowaways on Noah's Ark and woodworm, *Anobium punctatum*, has been a nuisance in the UK at least since Roman times. The Ancient Egyptians also had problems with pests – biscuit beetles, cigarette beetles and hide beetles have all been found associated with entombed burial materials.

It is clear that international trade has played an important part in the spread of many pests. The movement of large quantities of food by ship in the 19th century resulted in many pests such as biscuit beetles, flour beetles and spider beetles reaching a large number of countries. Trade in other commodities such as wool and timber helped to spread carpet beetles, clothes moths and woodborers.

Over the years, the change in emphasis has been from damage to personal belongings to include collections of objects specifically preserved for posterity.

In addition, the pest status of many insect species has changed over time. Many of these changes in pest status can be linked directly to changes in environmental conditions and in storage and display methods. There is now also much greater opportunity for insects to move from one country to another due to ease of travel which encourages much greater exchange of material between collections.

Despite the spread of pests, there are some distinct differences between pests found in different climatic areas. For example, the distribution of drywood and subterranean termites can be directly linked to the climatic conditions found in tropical and sub-tropical countries. The distribution of other pests may be much more local, but evidence from pest records in recent years shows that changes in pest status are occurring more rapidly. Some pests, such as the cigarette beetle, have become established in areas where they were previously unknown.

Insects are classified or divided up into groups according to their type of life cycle and the body forms of nymph, larvae or adult. Many types or orders of insects are not pests and need not concern us here. The major problems in museums and museum stores are caused by beetles (Coleoptera), moths (Lepidoptera), and termites (Isoptera). Local problems can also be caused by cockroaches (Dictyoptera), ants (Hymenoptera), silverfish (Thysanura) and booklice (Psocoptera).

What are insects?

Insect pests successfully exploit the situations which are provided in museums because features of their body form, breeding habits and life styles make them particularly well adapted to this role. An understanding of insect form and function is essential for successful IPM.

Insects are the largest section in the group of animals called Arthropods or 'jointed limbs'; they are also sometimes called 'hexapoda' because they have six legs. Although the term 'bug' is frequently used to describe any insect, it should only be applied to one group of insects called Hemiptera which includes bed bugs and plant bugs. Insects come in many shapes and sizes from butterflies to beetles and it would be impossible to describe all the different sorts of insect types and life styles in this book.

The most important and relevant characteristics of insects will be described in general and more detail is provided for specific pests.

Insect structure

The structural features of an adult insect are shown in Figure 1. The most important characteristics, and those which are very different from vertebrates like ourselves, are given below.

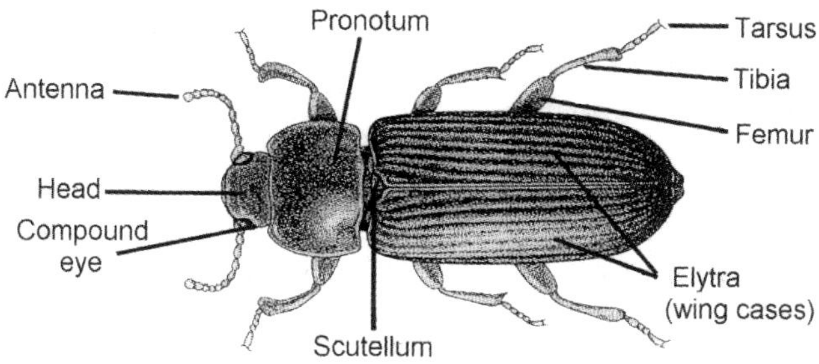

Figure 1 Mealworm beetle adult

Skeleton

The skeleton is on the **outside** and is a jointed box made of a hard substance called chitin. All the muscles are attached to the inside of this 'exoskeleton' which also holds and protects the body fluids and internal organs.

Legs and wings

Insects have three pairs of jointed legs; some moth larvae have extra pairs of false legs called prolegs. Many insect species have one or two pairs of membranous wings. The first pair of wings may be hardened into wing cases. The wing cases, or elytra, of beetles have a clear dividing line down the back of the adult.

Eyes and antennae

Insects have specialised sense organs and many species have multi-faceted 'compound' eyes which may not give clear images but are extremely sensitive to light and movement of objects. The most important sense organs of most insects are the antennae, long segmented filaments on the head which are provided with many sensory cells tuned to smell, taste, touch, humidity or heat.

Air

Like other animals, insects take in oxygen from the air and produce carbon dioxide as a waste product, but they do not have lungs. Instead they use a system of tubes called tracheae which branch from openings in the side of the insect called spiracles. They can close these off to prevent water escaping or poison gas getting in.

Nerves

Insects use nerves to transmit information both as input from sense organs and as output to muscles. Knowledge of how these nerves function helps in the design of effective pesticides.

Sex

Insects have sex organs or gonads when they are adults and frequently, but not always, there are two sexes, male and female. Female insects lay eggs, sometimes in very large numbers, which hatch into juvenile stages that do not have sex organs and other adult features like wings. The organisation of the body of an insect and its growth and development determine the pattern of the insect's life cycle.

Reproduction and growth

Because insects have an exoskeleton that is rigid, they cannot grow to reach the adult stage in the way that vertebrates do. Instead they use a system of periodic moulting to overcome this limitation. At intervals prior to moulting, an insect grows a new flexible skin under the old rigid one which is then split and shed. The insect body swells to expand the new skin which then sets hard. Insect growth therefore proceeds in a series of steps until the adult stage is reached and growth ceases. Adults often eat very little in comparison with the growing immature stages.

Insects use the series of growth steps in two completely different ways. The method which is thought to be the most primitive is called gradual or incomplete metamorphosis, shown in diagrammatic form in Figure 3. Cockroaches, silverfish and booklice all develop in this way and it can be seen that the developing stages or nymphs resemble small adults except that they do not have wings or sexual organs.

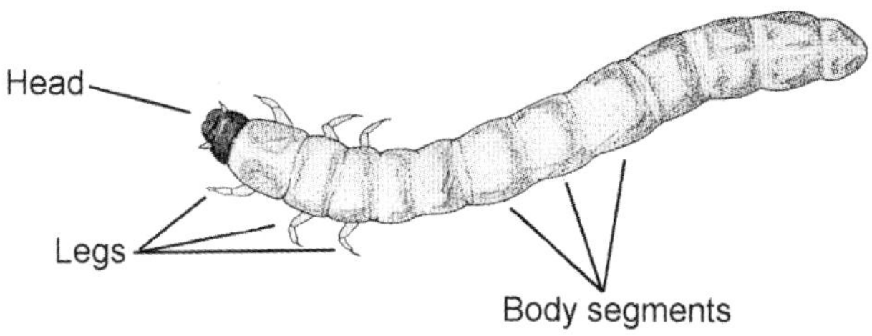

Figure 2 Mealworm beetle larva

The second method, called complete metamorphosis, is used by many insects including beetles, moths and flies and is shown in diagrammatic form in Figure 4. The developing stages or larvae are completely different from the adult in form and habit and often eat completely different food. The insect transforms from a grub-like larva to a sexually mature adult through an intermediate stage called the pupa. This appears to be inactive and relatively immobile but very complex changes are occurring internally.

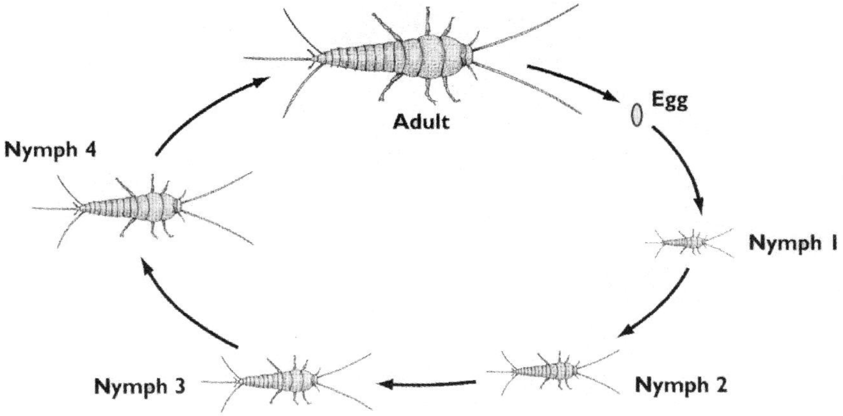

Figure 3 Life cycle of silverfish

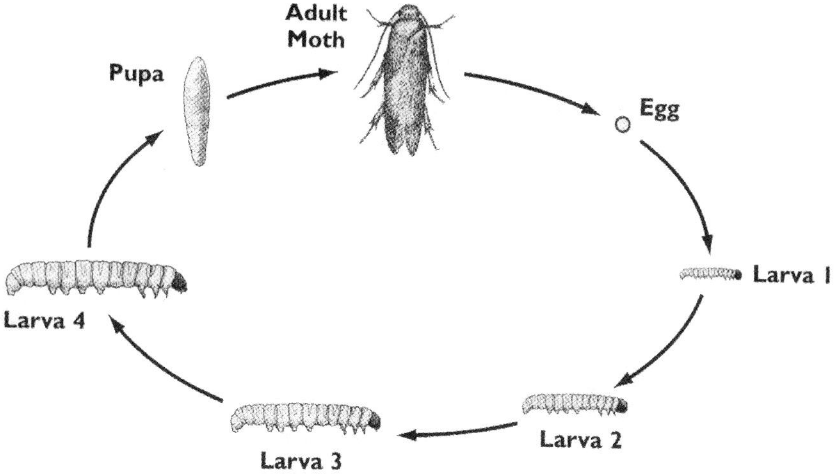

Figure 4 Life cycle of clothes moth

Environment, temperature, moisture and food

Insects are unable to regulate their own temperature and this can be of major importance in limiting or preventing problems. Body functions, and hence development and reproduction, are more rapid at higher temperatures and slow down as temperatures are reduced and eventually cease at low temperatures. Different species have different needs: some insects are adapted to tropical climates and others to temperate conditions. However, it is generally true that most insects will breed and develop rapidly above 25°C, will breed slowly between 15°and 20°C, and will not develop or breed at all below 10°C. Much lower temperatures are required to actually kill insects and this will be dealt with in Chapter 5.

Insects also require water, but whereas some species like silverfish require high levels of moisture in the air, others can obtain all the water they need from conversion of foodstuffs within the body. It is therefore impossible to make general statements about moisture needs of insects, although humidity control can reduce problems with some species, particularly those that are encouraged by supplementary mould growth.

All insects need food to grow and complete development but the life cycle illustrations show not only that different species may have very different food requirements, but that some species require different food at different stages of the life cycle. For example, a carpet beetle larva requires food of animal origin for growth and development and will eat wool, skin, dead insects and feathers etc. The adult beetle on the other hand eats very little but has a preference for pollen and nectar from certain wild flowers. Some species like the German cockroach have a very wide ranging diet and both adults and nymphs will eat nearly anything of animal or vegetable origin including excreta, confectionery and dried plant specimens. Other insects, such as the larvae of some of the woodboring beetles, will only develop in certain types of timber. Insects excrete digested food as small dry pellets (frass) or semi-liquid smears.

Given ideal conditions of abundant food, optimum temperature and RH and little disturbance, some species can complete development from egg to mature egg-laying female in as little as six weeks. This, coupled with the large number of offspring which can be produced from the eggs laid by one female, can give rise to enormous rates of increase and apparent population explosions – little wonder that museums and stores which provide an ideal environment for insects are often plagued with pests.

Major pest types

Because of the very large number of insect pests that occur in museums worldwide, it is impossible to deal with them all within the limitations of this book. Certain key species have been selected as being representative of the major pest types. For this introduction to pests and damage, the pests have been grouped according to the food they eat and by the damage they cause rather than by taxonomic features. An indication of the size of each species is shown by the life-sized image by each illustration and in addition, the length of the adult insect is given in the text. In all cases this should be taken only as a guide because of the large variation in size found in some species.

Accurate identification is essential and as there is only room to give relatively superficial entomological information, it is always advisable to have a provisional identification confirmed by an expert.

Pests which attack wool, fur, feathers and textiles

The main protein in these materials is keratin which is indigestible to many animals. Some beetle and moth pests have enzymes to break down keratin and they cause serious damage and destruction of materials of animal origin.

Carpet beetles (*Anthrenus* sp.)

The most common species found in Great Britain is the varied carpet beetle, *Anthrenus verbasci* (Fig. 5). There are a number of other species of *Anthrenus* which are similar in appearance and habits including the furniture carpet beetle, *Anthrenus flavipes* (common in the USA), and the Guernsey carpet beetle, *Anthrenus sarnicus* (Fig. 6), which is now well established in London and causing increasing problems in UK museum collections (Armes 1988). All adult *Anthrenus* are 2–3mm long and covered with grey and gold scales. Identification of the species is by the pattern and shape of these scales. Adult

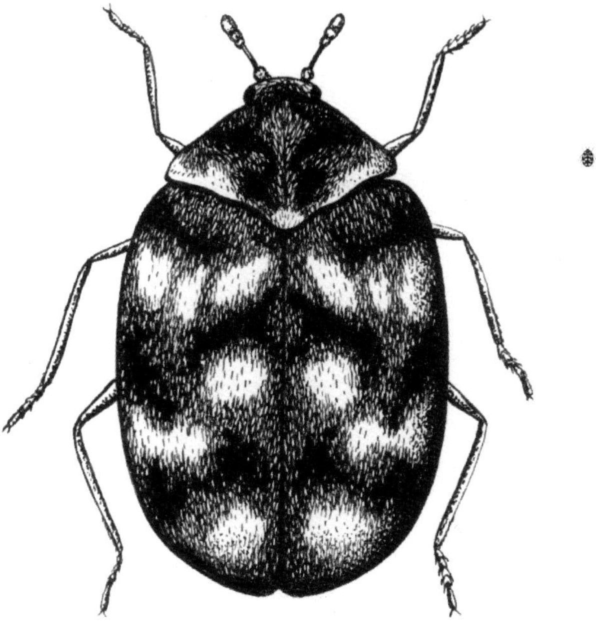

Figure 5 *Anthrenus verbasci adult*

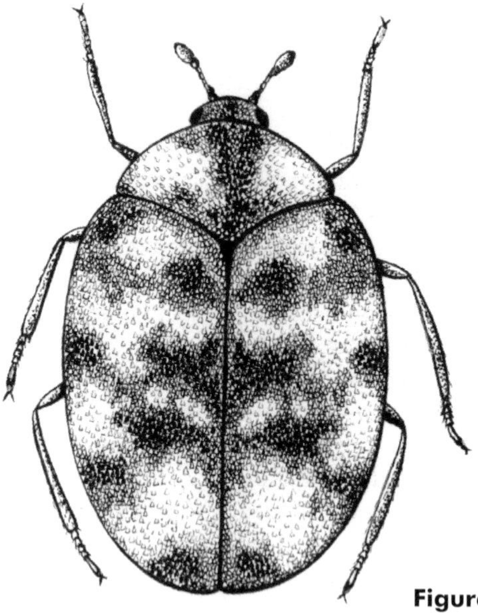

 Actual size

Figure 6 *Anthrenus sarnicus* adult

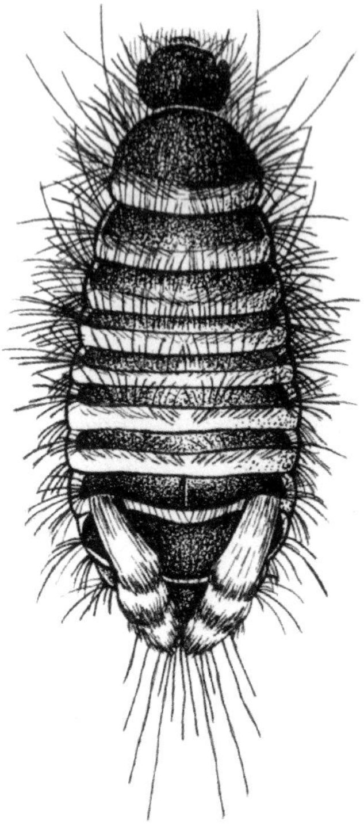

 Actual size

Figure 7 *Anthrenus* larva

beetles fly well in warm weather and may frequently be found on windowsills. In the UK they are found outside in late spring and early summer where they mate on flowers before returning indoors to lay batches of eggs secreted in cracks and crevices.

The eggs hatch into short, fat, hairy larvae up to 5mm long, often referred to as 'woolly bears' (Fig. 7). When the larvae first hatch they are extremely small, <1mm, and can gain entry to cases through very small cracks. As the larvae grow they leave empty hairy, cast skins which may blow around indicating the first signs of beetle attack. Larvae are voracious feeders and will rapidly demolish animal specimens, fur and feathers and woollen textiles. Clean cotton materials are not normally attacked. Carpet beetles are found in natural situations such as birds' nests and animal burrows, and in some museums and houses they have a direct route into the premises from birds' nests in attics and chimneys. Once established they can be difficult to eradicate because the larvae can forage widely and may take some years to complete development. Favourable conditions can encourage rapid development in less than a year causing high levels of damage.

Fur beetles (*Attagenus*)

The most common species in the UK is the two-spot carpet beetle *Attagenus pellio* (Fig. 8). There are other species of *Attagenus* including *Attagenus unicolor* (= *megatoma*) which is the most common species in the USA. The brown carpet beetle, *Attagenus smirnovi* (Fig. 9), is a pest in Eastern Europe and has now become established in London museums where it is known as the vodka beetle (Peacock 1979). *Attagenus* adults are variable in size, from 3mm to 6mm. *Attagenus pellio* is black with a distinctive white spot on each elytra. *Attagenus unicolor* is all black and *Attagenus smirnovi* has a black head and pronotum and brown elytra. They are active flyers and may be attracted to light sources. The larvae of *Attagenus* are covered in hairs and are longer and more torpedo-shaped than *Anthrenus* (Fig. 10). They usually have a tuft of bristles at the posterior end of the body. They are also very small when first hatched and shed their skins as they feed and grow to about 10mm. *Attagenus* are common in birds' nests and other animal debris and they are often found in the attics of old houses. They will damage insect collections and mammal and bird skins, and have been known to attack wool textiles. *Attagenus smirnovi* seems to require some starch in its diet and it has been found infesting flour and cereal residues. The life cycle of *A. pellio* is normally a year but *A. smirnovi* may complete its cycle more quickly at higher temperatures.

Hide and leather beetles (*Dermestes*)

Hide or leather beetles (*Dermestes*) as the name suggests, will attack leather and skins but they will not feed on tanned leather in good condition. The adults are black or dark brown and larger (6–10mm) than carpet beetles. The larvae are also dark brown, large and very hairy. *Dermestes maculatus* is black with

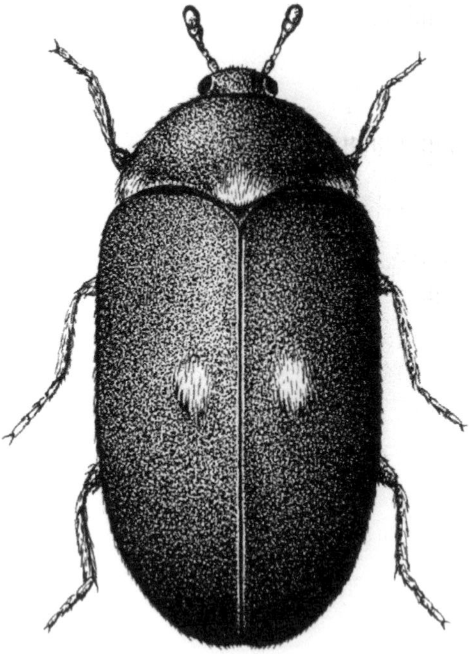

 Actual size

Figure 8 *Attagenus pellio* adult

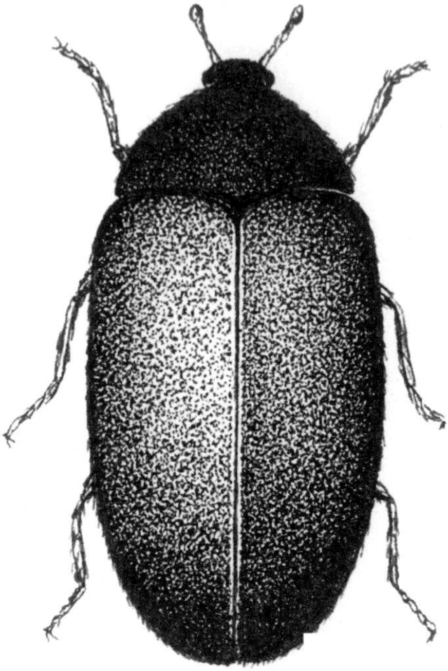

 Actual size

Figure 9 *Attagenus smirnovi* adult

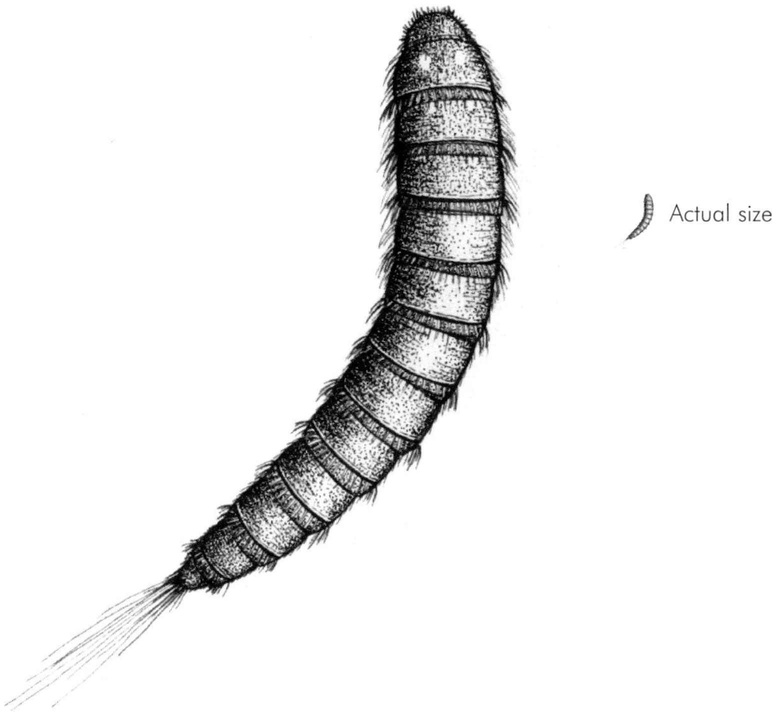

Figure 10 *Attagenus* larva

white hairs, is common in carcasses and only rarely found indoors. The Peruvian hide beetle, *Dermestes peruvianus*, has some yellow hairs and is becoming a common indoor pest (Figs 11 and 12). It is an active flier and is often found in urban areas in food refuse and catering establishments with poor hygiene. The presence of *Dermestes* larvae in a house or museum may be a sign that there is a dead rat or pigeon hidden away somewhere.

Some museums use cultures of *Dermestes* for cleaning animal carcasses to prepare skeletons. As they can cause havoc if they are allowed to spread into other areas, they should never be used in the main museum building.

Museum nuisance/American wasp beetles (*Reesa vespulae*)

This insect does not have a recognised common name and is very unusual in that it is parthenogenetic. This means that all the beetles are females which can lay fertile eggs without mating. As there are no male insects and one insect can start an infestation, the pest potential is greatly increased. *Reesa* is found in northern Europe and the USA where its natural home is insect nests. It has now become a major domestic pest in parts of Scandinavia. The adults are easily recognised as they are about 2–4mm long and have a yellowish band of hairs across the elytra (Fig. 13). The larvae are tubular and hairy and taper towards

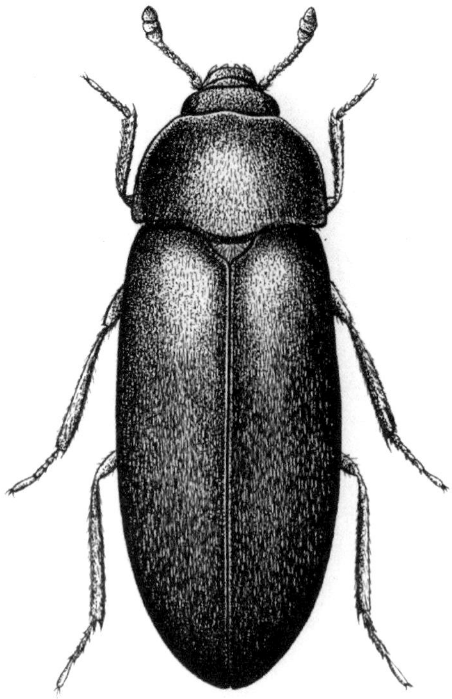

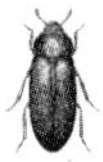

Actual size

Figure 11 *Dermestes peruvianus* adult

Actual size

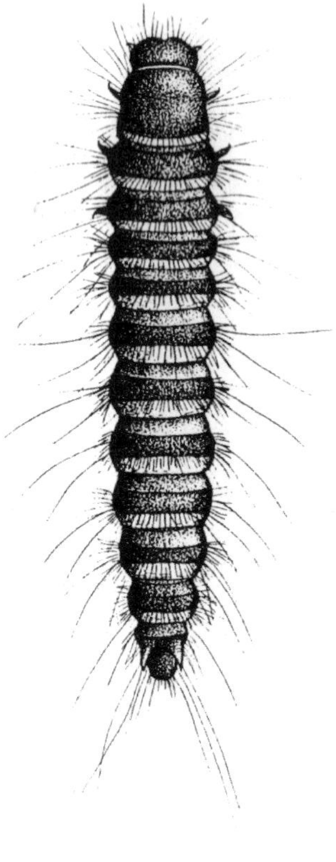

Figure 12 *Dermestes* larva

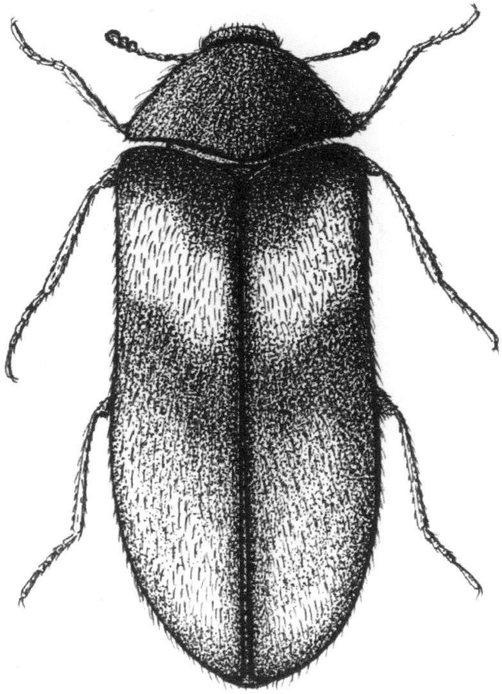

 Actual size

Figure 13 *Reesa adult*

the rear which has a tuft of hairs. *Reesa* was first found in the Natural History Museum in London in 1979 (Adams 1978) and it is now well established in the Entomology Department. It causes persistent and serious damage to insect specimens in isolated drawers. The larvae feed mainly on dried insects but they will also attack seeds and freeze-dried animals. In view of its persistence, museums should be very vigilant to prevent this pest becoming established.

Cabinet beetles (*Trogoderma angustum*)

This is the most recent pest introduction to the UK and does not yet have a common name. It was first recorded from the National Museum of Scotland, Edinburgh in 1998 where it was found damaging bird specimens (Shaw 1999). It was also established at this time in the herbarium of the Royal Botanic Gardens, Edinburgh. The adults are 2–4mm long and are quite distinctive with two bands of whitish hairs on the elytra (Fig. 14). It is a major pest in the Natural History Museum in Stockholm and appears to becoming more widespread in Europe. The first record from a museum in England was a specimen found in the herbarium of the Royal Botanic Gardens, Kew in 2000. Subsequently, a larva was also found in the herbarium. Museum staff should

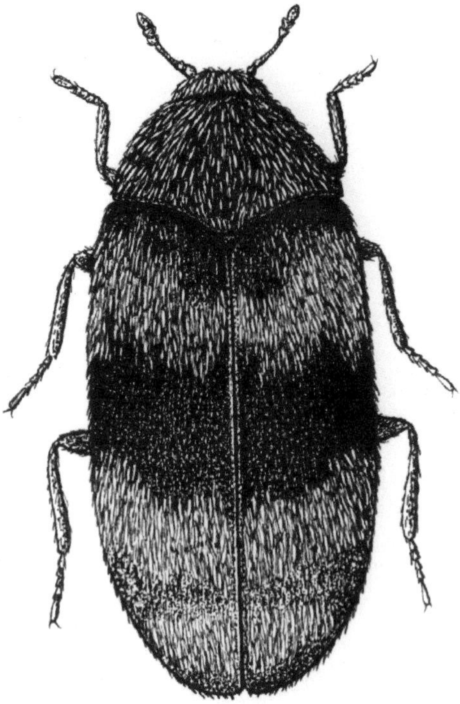

Figure 14 *Trogoderma angustum* adult

be aware of the potential of this pest to cause serious damage to plant and animal specimens.

Odd beetles (*Thylodrias contractus*)

This is called the odd beetle, and rightly so, as the male is 3–4mm long, has wing cases which remain partly open and is a very strange shape for a Dermestid beetle (Fig. 15). The female has no wings or wing cases and looks like a louse. The larvae are also very distinctive as they are short and fat with hairy bands and when disturbed, they curl up like a woodlouse (Fig. 16). *Thylodrias* is native to central Asia and has become established in some museums such as the Smithsonian in Washington. It has also become established as a pest in museums and houses in Finland and the larvae are known to eat dead insects and attack natural history specimens. The only known established breeding population of the odd beetle in the UK is in the herbarium in the Natural History Museum in London.

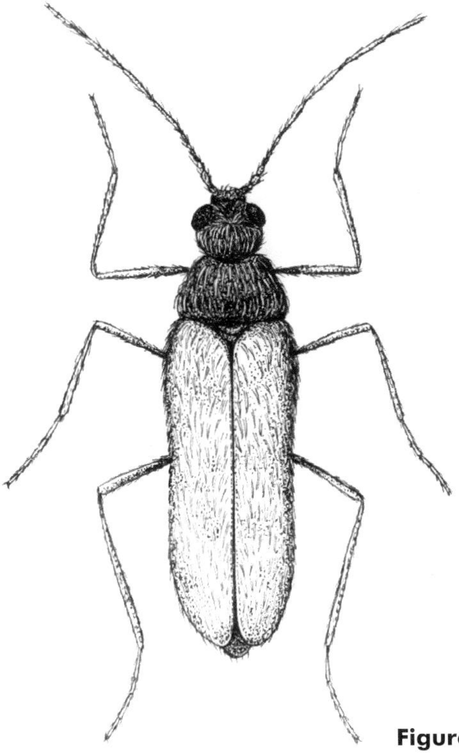

 Actual size

Figure 15 *Thylodrias contractus* adult

Actual size

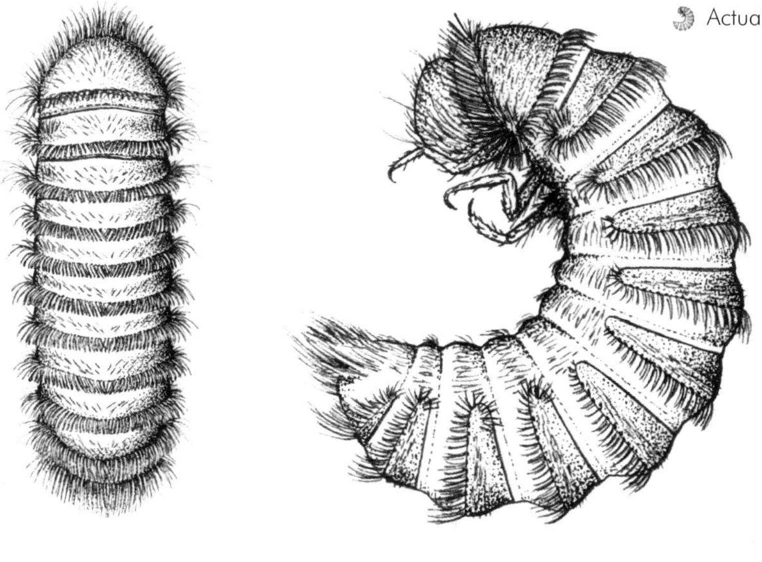

Figure 16 *Thylodrias contractus* larvae (top and side view)

Clothes moths (*Tinea* and *Tineola*)

A number of species of moth will attack and damage textiles and animal specimens. The case-bearing clothes moth *Tinea pellionella* and the common or webbing clothes moth *Tineola bisselliella* are two of the most important species with a wide distribution. They are small, dull moths, 5–7mm long, which scuttle around and only fly when it is warm. They shun the light, hide in dark areas and lay batches of eggs on fur, feathers, skins, wool or soiled silk. The larvae which hatch from the eggs spin silk webbing.

The case-bearing clothes moth larva *Tinea* spins a cocoon around itself leaving the front end open so that it can use its jaws and legs (Fig. 17). It then eats as it crosses the material carrying its case and leaving a trail of grazed textile or fur with some fragments of excreta or frass. The larva moults within the case and when fully grown it pupates within the cocoon and eventually the adult moth emerges to mate and lay eggs. The adults are greyish-fawn with two dark patches on each wing (Fig. 18).

The common or webbing clothes moth *Tineola*, has different habits from *Tinea*. Although the larva also spins silk, it leaves this as a tunnel or sheet of webbing across the attacked material under which it feeds (Fig. 19). Damage by this species is accompanied by copious webbing tubes or sheets, which frequently include large amounts of frass, and infestations appears far more 'messy' than the damage caused by *Tinea*. The adults are a shiny fawn colour without spots (Fig. 20).

Adult moths can fly in through windows or open doors and can also originate from birds' nests. One generation normally takes a year, but development can be more rapid if local conditions are suitable. The pelleted excreta or frass produced by the larvae of clothes moths is frequently mistaken for moth eggs. However, frass pellets are hard and opaque whereas moth eggs are very small, translucent and vulnerable to physical damage. Contrary to popular opinion, eggs will not remain dormant in textiles and then hatch many months later.

Textiles soiled with food or urine are preferentially attacked and a stained area may be more damaged than an adjacent clean one. Damage is also more concentrated in dark crevices and creases; for example behind lapels, in pockets or where carpets or textiles are folded. Clean cotton materials are not normally attacked.

Other moth species such as the white-shouldered house moth, *Endrosis sarcitrella*, and the brown clothes moth, *Hofmannophila pseudospretella,* are very common in old houses but very rarely cause damage to clean, dry textiles. The tapestry moth, *Trichophaga tapetzella* is very rare in the UK but in some countries it causes serious local damage to wool and silk textiles.

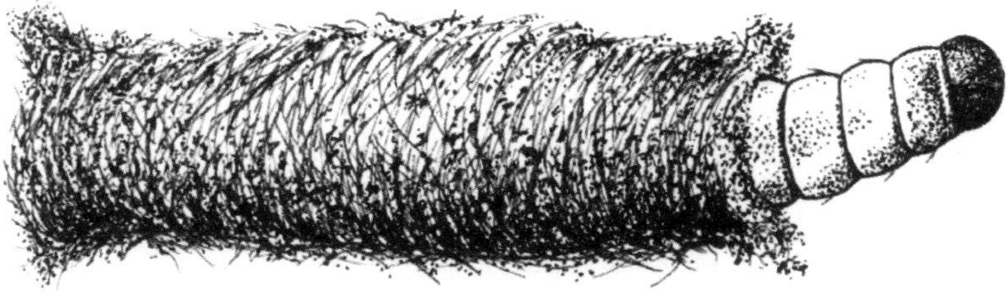

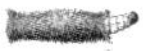 Actual size

Figure 17 *Tinea pellionella* larva

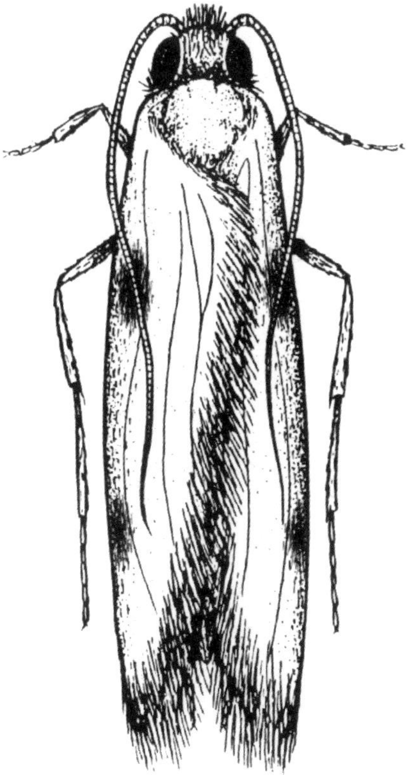

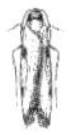 Actual size

Figure 18 *Tinea pellionella* adult

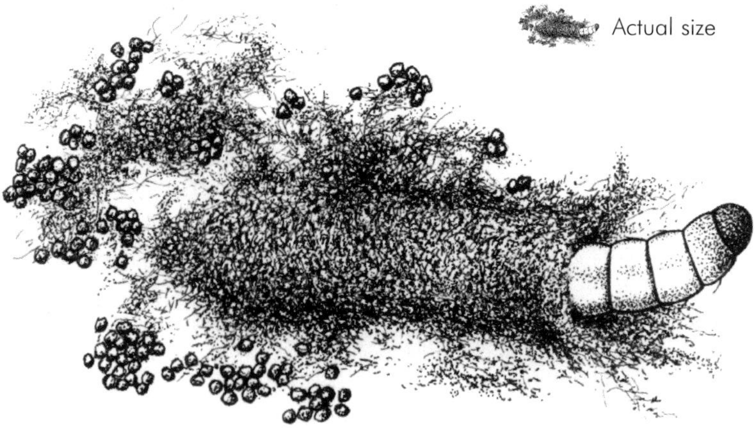

Figure 19 *Tineola bisselliella* larva

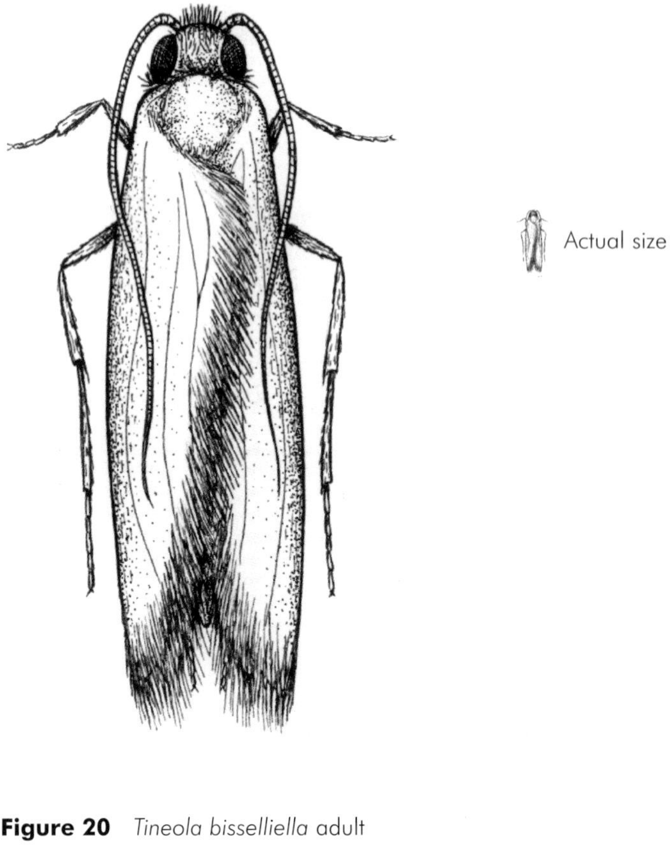

Figure 20 *Tineola bisselliella* adult

Pests which attack herbarium specimens and dried food

A number of species of beetles and moths can cause serious problems in houses and museums. Many of these pests also cause considerable problems for the food processing and storage industries.

Biscuit beetles or drugstore beetles (*Stegobium paniceum*)

The biscuit beetle or drugstore beetle, *Stegobium paniceum,* belongs to the Anobidae, the same family as the common furniture beetle or woodworm, *Anobium punctatum.* Unlike woodworm larvae, which only eat wood and cellulose, those of the biscuit beetle bore into hard dried vegetable material including biscuits, tobacco, nuts and dried plant specimens. Biscuit beetles will attack papier mâché which has a high starch content and there have been also problems of damage to freeze-dried animal specimens and fungi. Adults are reddish-brown and 2–3mm long (Figs 21 and 28) and when it is warm they are very active and will fly to light sources. The larvae are white and curved and they will tunnel through hard materials. Development time depends upon food and temperature – in nutritious food at 30°C it can be as short as five weeks. At lower temperatures, the life cycle may take six months to a year. When the adults emerge they leave neat, round exit holes in objects. Biscuit

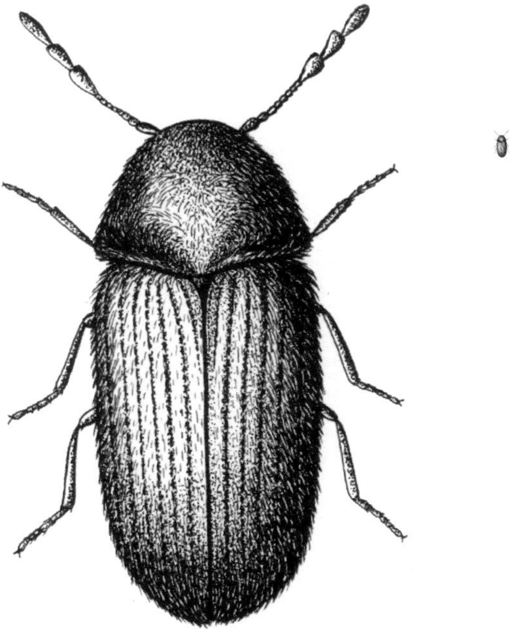

Figure 21 *Stegobium paniceum* adult

beetles have an amazing ability to survive and breed on drugs and spices, some of which are extremely toxic to other animals. A large number of biscuit beetles can be produced from a relatively small food source. For example, hundreds of beetles were found in one museum store where they had been breeding in a few old loose-grain mouse baits.

Cigarette beetles or tobacco beetles (*Lasioderma serricorne*)

Cigarette beetle adults are similar in size and appearance to the biscuit beetle but they have quite different antennae and no lines on the elytra (Figs 22 and 28). The white larvae are also very similar to *Stegobium* and make similar tunnels and exit holes. They attack a wide range of dried plant material, including tobacco, and they will also eat animal protein such as freeze-dried animal specimens. *Lasioderma* is a common pest in tropical and semi-tropical climates where it replaces *Stegobium*. It is occasionally found in the UK in heated buildings and has recently caused problems in the herbarium at Kew.

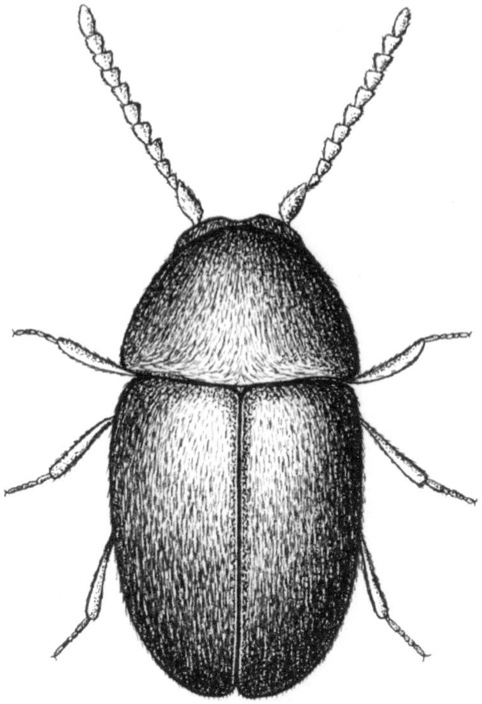

Figure 22 *Lasioderma serricorne adult*

General detritus feeders, mould feeders and scavengers

There are many insect types and species in this group. They generally cause nuisance and low levels of damage but occasionally populations of insects can build up to levels sufficient to cause serious damage.

Spider beetles

Spider beetles are common in birds' nests and general debris in attics, basements and stores where they will feed on a wide range of vegetable and animal detritus. The adults are 3–5 mm, hairy and superficially spider-like; they are generally slow-moving and may breed slowly. The Australian spider beetle, *Ptinus tectus*, has a brown hairy body and is common in many temperate countries (Fig. 23). The golden spider beetle, *Niptus hololeucus*, has a round body covered in shiny golden hairs and is becoming more common in historic houses. Other species include the white-marked spider beetle, *Ptinus fur*, and the shiny or hump-backed spider beetle, *Gibbium psylloides*. The larvae are similar in appearance to those of the biscuit beetle and they will also bore holes

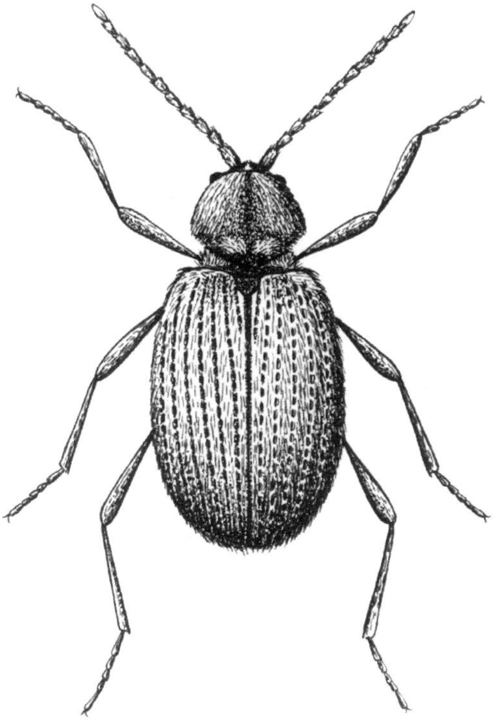

Actual size

Figure 23 *Ptinus tectus adult*

in hard materials before pupating in a globular silk cocoon. They will often damage paper and wood near infested food. Spider beetles will tolerate lower temperatures than many other beetles and some species can complete development below 10°C. At higher temperatures, two generations a year may be produced and damage may be severe. They have been known to damage insect collections, animal skins, dried plants, books and textiles.

Mealworm (*Tenebrio molitor*)

These large (10–15mm), black, slow-moving beetles are frequently found wandering around in attics and upstairs rooms (see Fig. 1). They have large, leathery, yellowish larvae (Fig. 2) which live in old organic detritus, such as birds' nests. They do not cause damage to objects but are indicators of poor hygiene. They are sometimes confused with black ground beetles which are outdoor active predators usually found near doors on the ground floor.

Booklice

There are a number of different species of booklice found in museums, houses and stores including, *Liposcelis*, *Lepinotus* and *Trogium*. *Liposcelis bostrychophila* is one of the most common species found in heated buildings and is also one of the most troublesome insects. The adults are wingless and very small, less than 1mm (Fig. 24). Booklice develop through a series of nymphal stages

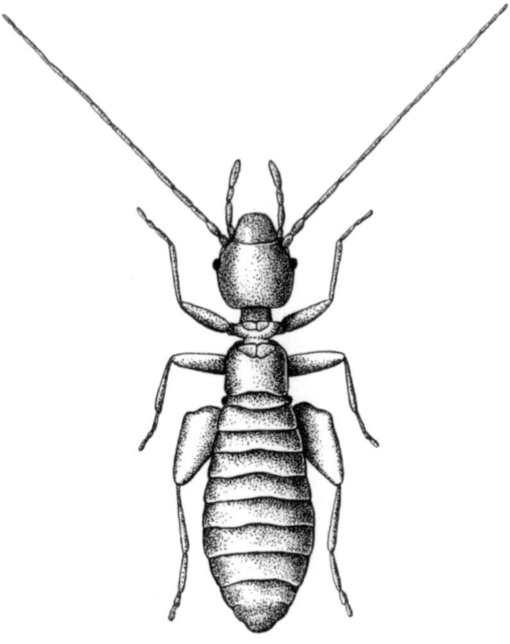

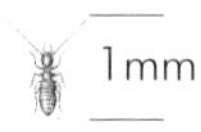

Figure 24 *Liposcelis* adult

which feed on microscopic moulds on a range of substrates including flour, paper and cardboard. They may also damage insect collections. Because there are no males and the females do not require fertilisation to lay eggs, populations of *Liposcelis* can increase very rapidly at temperatures over 25°C and give rise to apparent population explosions.

Although direct damage may not be serious, large numbers of booklice will graze the surface of books and papers. In addition, squashed bodies will stain materials and may encourage moulds. While these insects thrive at high RH which encourages mould growth, *Liposcelis* can tolerate drier conditions than some other booklice and will develop when the RH is 60%. Booklice can be a particular problem in books and manuscripts but if these are kept in clean, dry and cool conditions, they are not at risk.

Silverfish

Silverfish are always associated with damp conditions and they generally require localised RH above 75–80% to breed and multiply. They are primitive, scaly, wingless insects (10–15mm) with forked bristles at the tail end (Fig. 25). Silverfish are general scavengers and feed on microscopic moulds on a wide range of foods including glue and ink on paper. *Lepisma saccharina* has been known to damage wallpaper, labels, postage stamps and paper currency. Silverfish damage can be recognised by the ragged, scraped areas and irregular holes in paper. They are useful indicators of humidity and if silverfish are found in numbers in an apparently dry room, it is likely that there are high humidity micro-environments nearby.

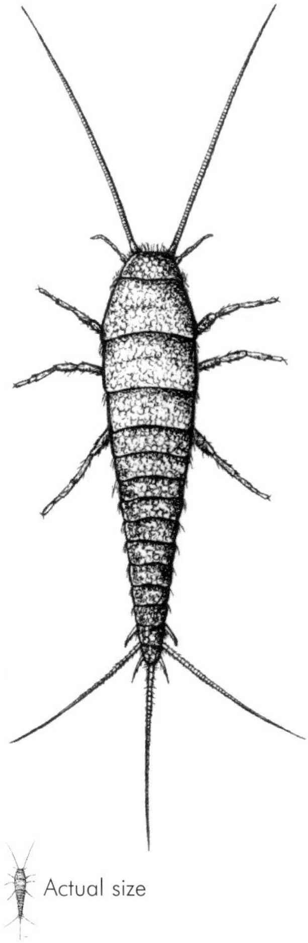

Figure 25 Lepsima adult

Cockroaches

Worldwide, cockroaches are one of the most serious pests associated with man and buildings, and cockroach problems sometimes spread to museums and museum stores. All cockroach pests are tropical in origin and require relatively high temperatures. Many species also require locally high RH or some free water. They are extremely active and the adults of some species will fly. They are also relatively large but as they frequent heating ducts, wall cavities and other dead spaces, the centres of infestation may be hidden. Although tropical in origin, German cockroaches *Blattella germanica* (10–15mm) and Oriental cockroaches *Blatta orientalis* (20–30mm) are very common in heated buildings in countries with temperate climates. The American cockroach *Periplaneta americana* (30–40mm) is probably the most abundant insect in countries with tropical and semi-tropical climates. All the pest species lay eggs in cases called oothecae and tiny nymphs hatch from these.

In temperate countries, infestations of cockroaches may spread from restaurant or catering areas. They will eat a wide range of human food and they will also feed in toilets and drains. As they can carry a range of diseases and cause allergic reaction, cockroach infestation can present a health hazard to staff. Cockroaches are active at night and the size of infestations can be underestimated if inspection is only carried out in daylight. Infestations may be difficult to eradicate even with the use of baits and persistent pesticides and successful cockroach control usually requires close co-operation between museum staff and an outside pest control contractor.

Ants

Some species of ants may invade museum premises at certain times of the year. For example, local problems can be caused by workers of the black garden ant *Lasius niger* which leave their outside colonies and forage in buildings for food material. They will cause little damage but can be annoying to staff. Problems are usually cured by preventing entry and by improvements in hygiene in areas where food may be left out or spilt. More serious problems can be caused by tropical species such as Pharaoh's ant, *Monomorium pharaonis*. This is a very serious pest in hospitals and restaurants and will sometimes colonise adjacent buildings. Control of Pharaoh's ants usually requires the services of a pest control contractor.

Woodlice

These are not insects but belong to a group called the Crustacea which also includes shrimps and crabs. They have many legs and a series of overlapping plates covering the top of the body. There a number of species which come into buildings; greyish-brown in colour and ranging in size from a few millimetres to 15mm. They live in damp, rotting vegetation and wood and cannot survive for long in dry conditions. They are often found in basements or near doors and windows where they have wandered in from damper outdoor environments. Most soon die of desiccation without causing any damage. If there is a persistent problem of live woodlice, then the area should be investigated as there may be local high RH and rotting wood.

Woodborers

Many different sorts of insects will attack timber during its life from a standing tree to a rotting log. Often pests are specific to particular types of wood and whether it is living, dried or decaying. Generally, the adult insects cause little damage – it is the larvae which cause damage ranging from one emergence hole in an object to the complete destruction of roof beams.

Many of the most destructive pests are beetles, however, wood wasps and some species of ants can also cause damage. Termites are extremely serious pests in some countries and will be dealt with separately (see p. 42).

It is important to distinguish between those insects which will only start development in living timber but may subsequently emerge from objects made from infested wood, and those which infest cut seasoned timber. Although the former are less important they can occasionally cause severe damage. Large longhorn beetles, for example, can leave an exit hole 5cm in diameter, but these insects present no further hazard to stored artefacts or the building as the females will only lay their eggs on trees or recently felled timber.

The pests of seasoned timber present the greatest hazard, particularly beetles which can emerge from the wood, mate and immediately lay their eggs in wooden objects, wooden racking, furniture or even in the store structure.

It is important with all woodborers to distinguish between wood which contains active infestation and wood with exit holes which is damaged, but no longer contains living eggs, larvae or adults.

Furniture beetles (*Anobium punctatum*)

The common furniture beetle or woodworm is widespread in most temperate countries where it infests buildings, furniture and wooden objects. The dark-brown beetles are 3–5mm long (Figs 26 and 28) and emerge during the spring and summer months by gnawing circular exit holes (1.5–2mm diameter) in the wood. As they fly or crawl away, wood dust and frass may fall out of the hole. This leaves the characteristic sign of active beetle infestation – small piles of fresh bore dust. The frass feels gritty and is the shape of wheat grains when seen under the microscope (see Fig. 33). After the beetles have mated, the female lays eggs in crevices in the wood particularly in end grain, unplaned timber or old exit holes. The eggs hatch in a few weeks, the larvae tunnel into the wood and as they gnaw, the size of the tunnel increases (Fig. 27). The larvae take 2–3 years to complete development to adult depending upon the type of wood and the conditions of temperature and moisture content. *Anobium* infestations will survive in cool, damp conditions but do not thrive in dry conditions with RH below 55% (Hansen 1998). Adults will emerge from infested wood moved from damper to drier conditions, but the eggs and young larvae will not survive. This means that objects and structural wood in centrally heated buildings are very unlikely to support woodworm infestation. Outbreaks of woodworm activity are usually confined to outbuildings or areas where there are leaks, condensation and poor air circulation.

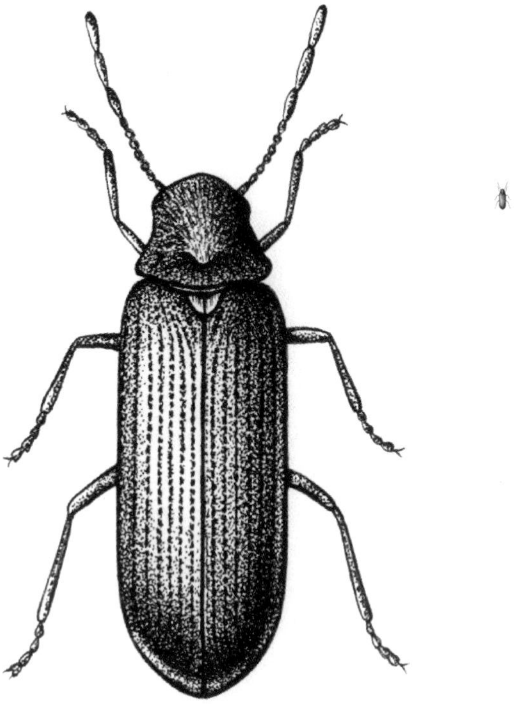

Figure 26 *Anobium punctatum* adult

Anobium larvae will attack most wood except sound heartwood, but they prefer starchy hardwood and softwood. Plywood from the first half of the 20th century is particularly susceptible because of the added protein from casein and blood albumen, and it can be severely damaged. Woodworm larvae will even develop in books and wood pulp paper when the leaves are compressed, which is why the larvae are sometimes called 'bookworms'. They will rarely attack books that are clean and dry and arranged on shelves although the infestation may spread from infested wooden shelving.

Death watch beetles (*Xestobium rufovillosum*)

The death watch beetle is one of the most well-known pests of timber. However, it is unlikely to cause problems in museums situated in new buildings because it is confined to the old timbers found in historic buildings. *Xestobium* larvae attack hardwoods, particularly oak and elm; softwood is usually only attacked when it is in direct contact with infested hardwood. Damage to old timber can be very severe and result in total failure of load-bearing beams. The adults are brown with yellow scales and are quite large (6–9mm) (Fig. 29). The name 'death watch beetle' comes from the tapping sound made by the beetles

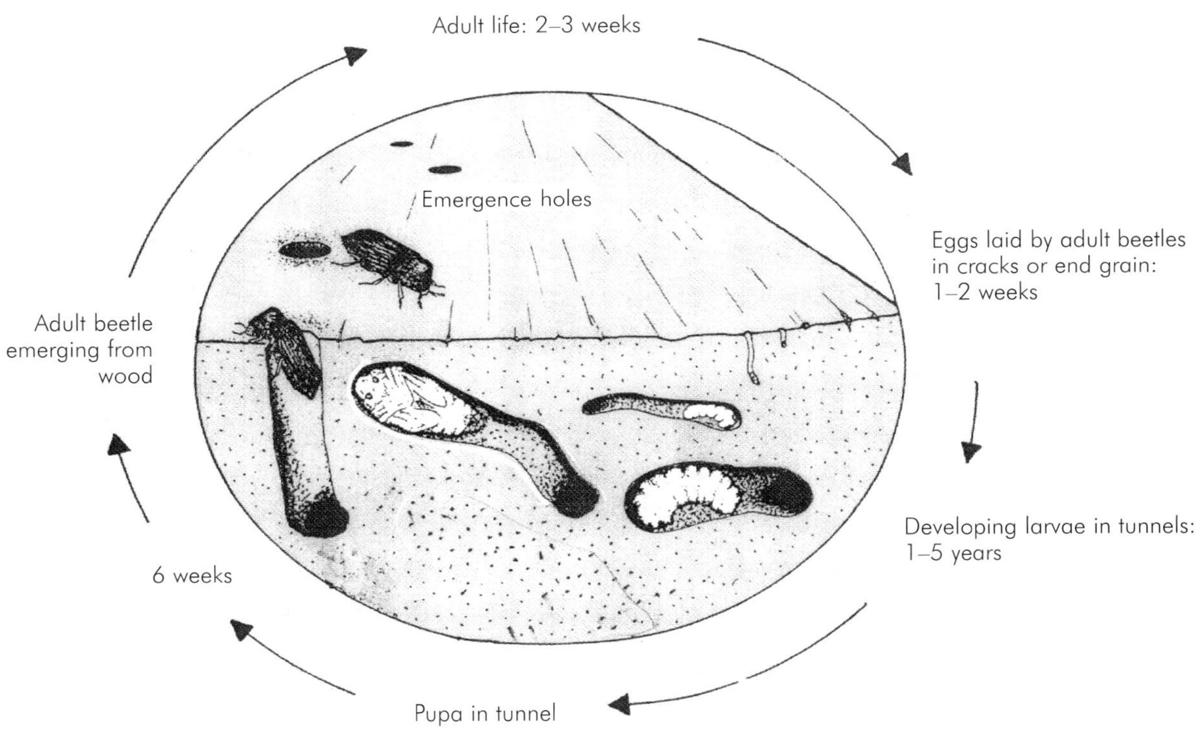

Figure 27 *Anobium* life cycle (based on Bravery *et al.* 1987)

(a)

(b)

(c)

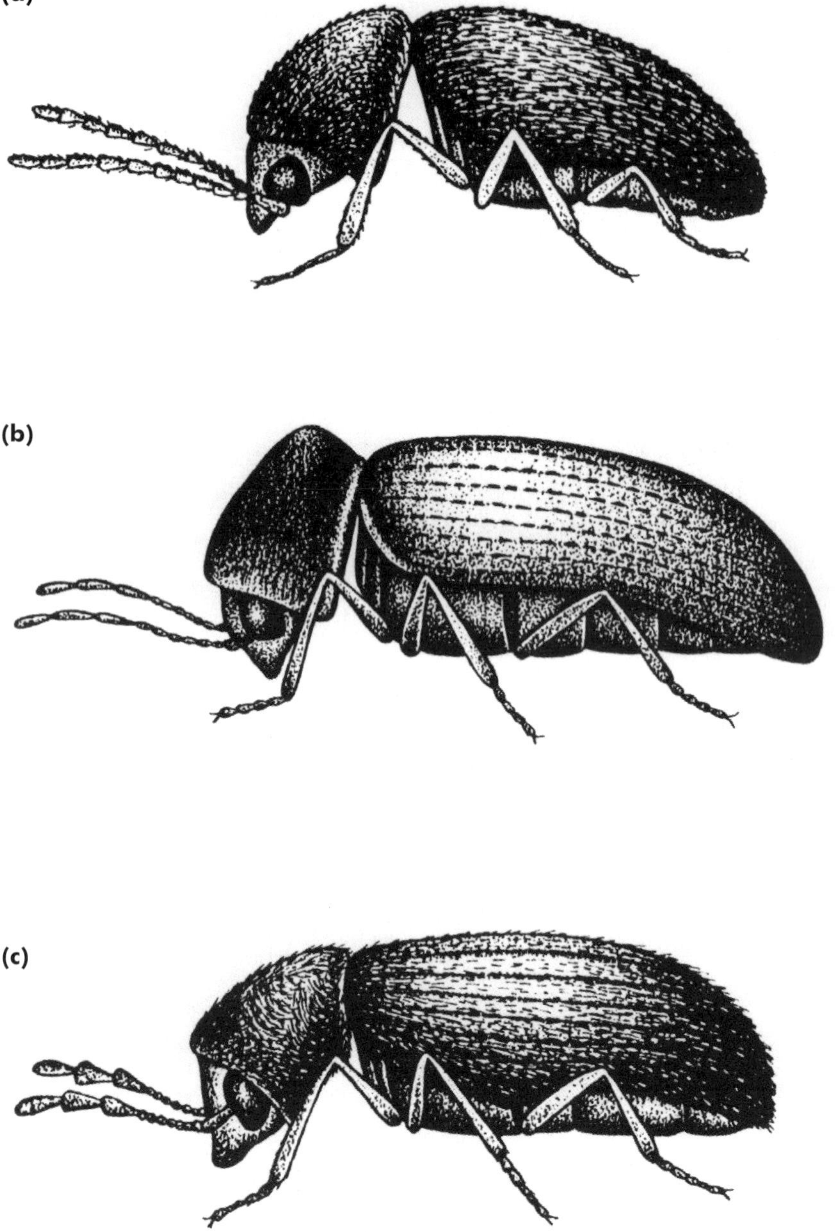

Figure 28 Anobiid beetles, side views: (a) *Lasioderma serricorne*; (b) *Anobium punctatum*; (c) *Stegobium paniceum*

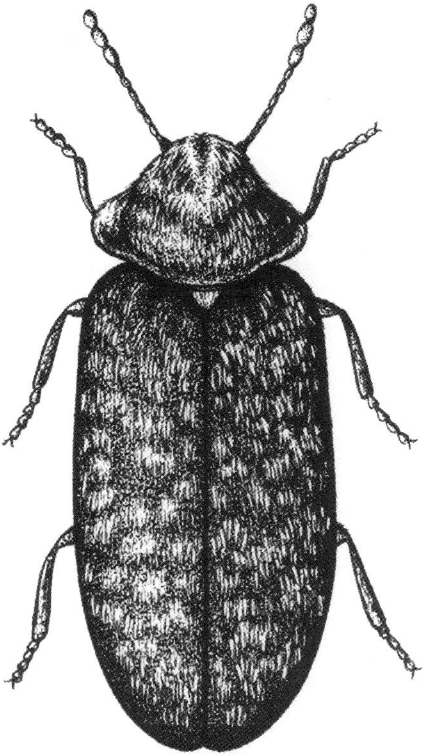

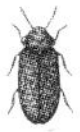

Actual size

Figure 29 *Xestobium rufovillosum* adult

Actual size

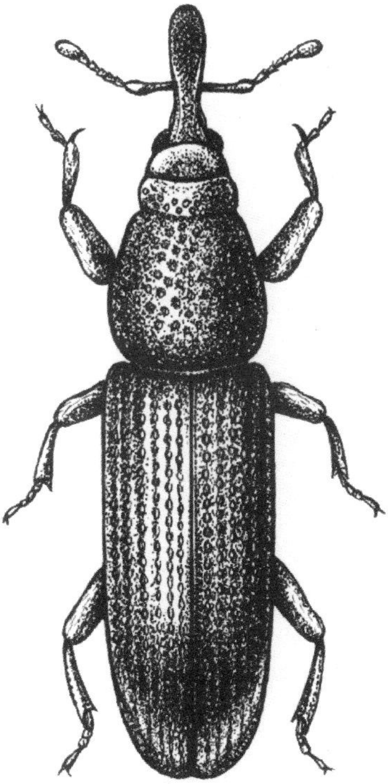

Figure 30 *Euophyrum* adult

which can be heard in warm weather between March and June. The larvae take 5–10 years to develop and adult exit holes are large and up to 3mm in diameter. The frass which falls out of infested timber is very coarse, round pellets (see Fig. 33).

Staff in old houses and museums in old buildings with oak timbers should be vigilant for signs of attack in bearing ends of roof beams which may be in contact with damp walls. If activity is suspected it is usually advisable to seek expert advice. When parts of an old building which have been attacked by *Xestobium* are placed in a museum they should be checked for signs of active infestation. Insects can emerge after many years and although they are unlikely to damage other objects, the beetles could spread into old timber in the structure.

Wood weevils

These are small, 2–3mm, cylindrical, black beetles with a long snout called a rostrum (Fig. 30). There are two similar species, *Euophryum confine* and *Pentarthrum huttoni* which are only found in damp wood. The exit holes are very small (1mm) and the surface of attacked wood often looks eroded and irregular. They are usually found in damp basements and are a good indicator of high moisture levels in wood.

Longhorn beetles

The most common in buildings is the house longhorn beetle, *Hylotrupes bajulus*. This is a large (12–15mm) beetle with long antennae. Its original habitat is old dry conifers but it is now a major pest in buildings in some countries. The larvae bore mainly in the sapwood, creating tunnels of increasing size and taking 3–5 years to complete development. When fully grown, the larva bores outwards towards the surface and leaves a thin layer of wood remaining before it pupates in a special chamber deeper in the wood. When the beetle emerges from the timber, wood chips, dust and cylindrical frass (see Fig. 33) fall out of the large, oval 6mm × 3mm, exit holes. The adults require temperatures of at least 25°C in the summer months and in the UK they are confined to a very restricted area in southeast England. Because of the larval habit of leaving a thin layer of protective wood on the surface, severe structural damage can sometimes occur before the infestation is detected. Furniture and other objects recently imported into the UK may carry larvae inside which can be heard scratching the wood as they eat.

Powder post beetles (*Lyctus*)

Many pests can be recognised by the characteristic dust and frass which falls from larval tunnels. Infestation by powder post beetle *Lyctus brunneus* is easily diagnosed by the presence of very fine talc-like powder around damaged objects. The larvae develop in deciduous hardwoods such as oak, tropical hardwoods and bamboo, and after about one year's development, the adults

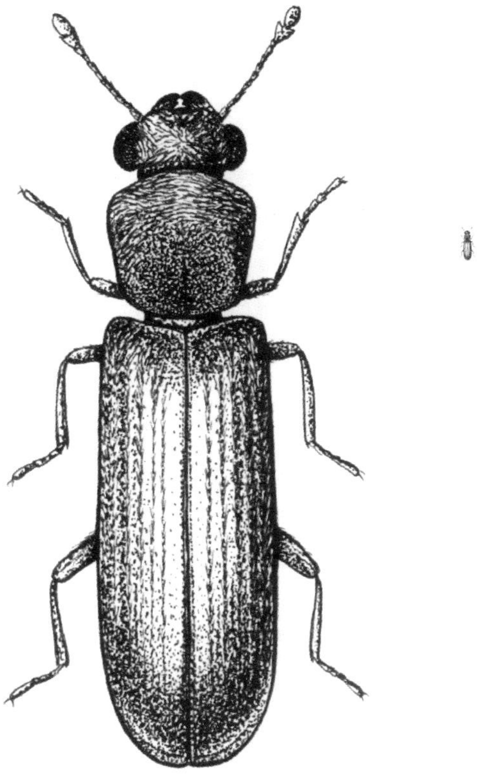

Figure 31 *Lyctus brunneus* adult

leave 1.5–2mm diameter exit holes. The adults are long (5–6mm), flat, reddish-brown beetles (Fig. 31). Severe infestations will reduce wood to a powdery mass surrounded by a paper-thin shell. They usually only attack relatively fresh wood with a high starch and sugar content. Established infestations are rare in the UK but they are often introduced on wood and furniture from warmer countries. Old wooden objects and structural timber in museums and houses are not at risk.

Bostrychid borers

The bostrychids are a large group of mainly tropical beetles which attack hardwoods and bamboo. Frass is powdery and damage is similar to that caused by powder post beetles and they may destroy the internal structure of the wood. The adults leave exit holes from 1mm to 5mm in diameter depending upon the species. The auger borer, *Sinoxylon*, and bamboo borer, *Dinoderus*, are very common pests of starchy wood and bamboo in tropical countries.

Wharfborers (*Nacerdes melanura*)

Waterlogged timber may be infested by larvae of the wharfborer. The adult beetle is relatively large (8–12mm) and flies actively. The white larvae are also large with distinct protuberances and they only thrive and tunnel in wood which is saturated with water. Damage to timber building supports in basements has been found in addition to the more typical damage to wharf pilings and ships' timbers.

Parasitic wasps

Timber which is infested with beetles may also attract some parasitic wasps. Small bright metallic green chalcid wasps (2mm) will lay eggs in *Anobium* larvae. The wasp larvae will then develop in the parasitised *Anobium* until the adult wasps emerge through separate and very small holes in the wood.

Ants

Most of the ant species which cause problems are associated with decaying timber, but carpenter ants, *Camponotus* sp., will gnaw tunnels to build nests in soft spring wood. The ants do not attack hard winter wood and so the tunnels and damage follows the growth rings. Structural timber of buildings and certain large artefacts may be damaged.

Termites

Termites, commonly known as 'white ants', are not ants but members of a group called the Isoptera. They are the world's most serious and destructive pests of structural timber. In many museums, termite infestation of the buildings has spread to display and storage furniture and objects causing serious damage. Termites are not a problem in cool temperate countries and are absent from Canada, Northern Europe, the USSR and northern Asia. The most severe problems are encountered in countries with warm temperate and tropical climates.

There are many different species of termites but they are generally divided into three types with distinctly different life styles – drywood, dampwood and subterranean. All three types can cause problems in museums and stores. Subterranean termites cause the greatest damage to buildings whereas drywood termites cause the greatest damage to artefacts (Edwards and Mill 1986).

Termites, like ants, live in colonies which may have millions of individuals with a complex social structure. There are different types or 'castes' in this structure which includes workers (Fig. 32), soldiers, kings and queens, plus the developing eggs, larvae and nymphs. The reproductive cycle of many termites involves the production of large numbers of winged reproductives called alates which swarm from the colony and then lose their wings before establishing new colonies.

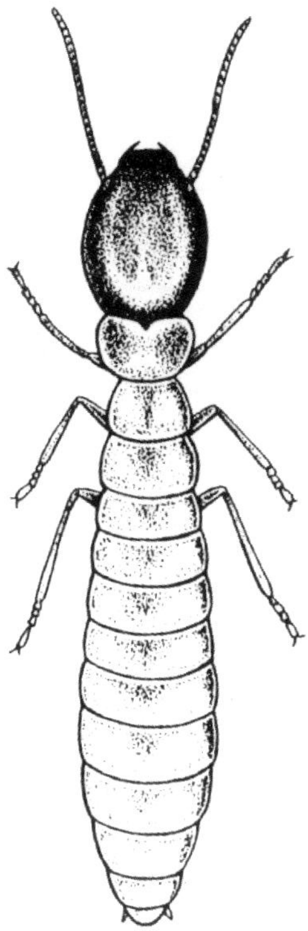

Actual size

Figure 32 Termite worker

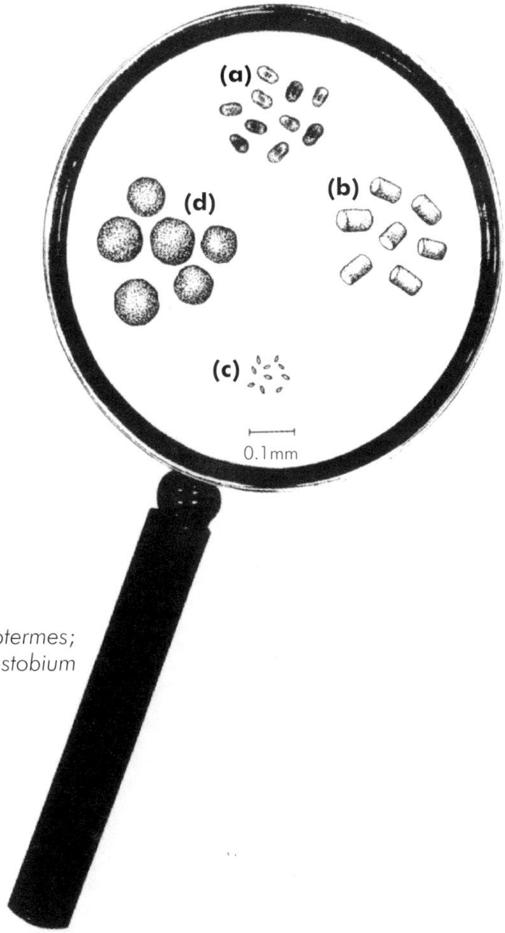

Figure 33 Insect frass: (a) *Crytotermes*; (b) *Hylotrupes*; (c) *Anobium*; (d) *Xestobium*

Drywood termites

Drywood termites, which include the species *Cryptotermes* and *Kalotermes*, bore galleries in many directions and tunnels, frequently across the grain of the wood. Channels are connected by narrow passages that are clean and smooth. Although some faecal pellets or frass are scattered in the passages, they are often stored in chambers which are sealed off. Drywood termites can be identified by these pellets which are the shape of poppy seeds (Fig. 33).

The termites may bore small round holes in the surface of the wood and eject the frass through these 'toilet holes'. Timber and wooden objects may be extensively damaged before infestation is noticed because a thin layer of undamaged wood may be left on the outside. The nests and colonies are constructed entirely within timber and they do not forage widely from infested material. Drywood termites will also attack and destroy other cellulosic material including books.

Dampwood termites

Dampwood termites including *Zootermopsis* sp. are usually associated with wood which is already being attacked by fungus and their presence indicates advanced decay. The tunnels and galleries are similar to those made by drywood termites but the interior walls are not so clean and may be littered with pellets.

Subterranean termites

Subterranean termites need to live in contact with soil and some species build the classic 'termite cities' with colonies in huge mounds or spires of soil. Colonies of some species, including *Reticulitermes*, *Coptotermes* and *Macrotermes*, will spread from the natural environment of soil and trees to the woodwork of buildings.

Many species require fungi in their diet which is produced on decaying wood within the nest. They need to maintain high levels of moisture in the colony for the development of fungi to prevent the desiccation of the nymphs and workers. This leads to the characteristic tube-building habitat of subterranean termites. These tubes, which may be some metres in length, are constructed of soil and faecal material to protect the termites as they pass between nests in soil to the food sources of wood or other cellulosic material. They will line their galleries with a mud layer that conceals holes and covers wood which they are excavating. This habit, together with the presence of connecting tubes, provide the main clues to distinguish colonies of subterranean termites from those of drywood termites with their relatively clean galleries. A further diagnostic feature of damage is that subterranean termite galleries are restricted to softwood and do not cross growth rings (Anon. 1999).

As their name implies, subterranean termites are found near or below ground level and seldom spread above the lower floor; this means that basement areas are particularly at risk. In addition to damage to museum objects, the enormous amount of timber which can be consumed by subterranean termites can be sufficiently severe to cause structural collapse of buildings.

The eradication of termites from museums and their subsequent exclusion can be extremely difficult and usually requires the services of a specialist pest control contractor.

Mites

Mites are not insects but are more closely related to ticks and spiders. They are usually extremely small (0.1mm) and difficult to see with the naked eye. Adult mites have an undivided body and four pairs of legs. The flour mite, *Acarus siro*, and house mite, *Glycyphagus domesticus*, have long hairs on the body and legs. They cannot survive in dry conditions and generally only cause problems when localised RH exceeds 70%. They will continue to breed and multiply at relatively low temperatures, e.g. 10°C, and infestation can occur in cool, damp conditions. Damage to objects may not be severe, but mite infestation indicates bad storage conditions. In addition, some mite species can cause serious allergic reactions in staff who handle infested material.

Although they are not strictly museum or even indoor pests, red clover mites, *Bryobia praetiosa*, can cause serious problems when they invade buildings. Clover mites do not cause any direct damage but large numbers can be extremely annoying. When the bodies become squashed on objects or between pages of books or documents, the body contents can cause severe red staining. The mites feed on grass and other plants, especially pea and clover, and in summer they seek shelter in nearby buildings. Large numbers may invade museums and houses, particularly on south-facing walls, and because they are 1mm or less in size, they can get through tiny cracks around doors, windows and pipework.

CHAPTER THREE

Insect detection and monitoring

Where do insects come from?

Insects are not spontaneously generated and therefore, if infestation develops, it must have been acquired from some other source.

The most obvious way for insects to gain access to houses and museums is through doors and windows from the outside world. Carpet beetles and furniture beetles fly readily in warm summer months and only one fertilised female is enough to start an infestation. In general, the warmer the climate, the more insects fly and the greater the risk from invasion.

Most old buildings will harbour some pests and it is quite common to find pests living within a building for many years without apparently causing a problem. It is not until their numbers increase and they move into displays or stored material that serious damage occurs. Birds' nests and dead animals in attics are the natural home of many pests where they live quite happily on feathers, hairs and excreta. When insect numbers exceed the food available, they will spread out in search of other breeding sites.

New material which is already infested can be introduced into collections and the insects may remain undetected until a serious problem develops. This can be a particular problem with woodboring beetles, some of which take years to develop. Material sent out for display or on loan may be returned together with infestation which has been acquired while it was away. There are documented cases of costumes returning to museums infested with tiny clothes moth larvae which had hatched from eggs laid by moths in the clothes when they were on open display. Packing cases and packing materials may also be a source of insect infestation.

How to find insects, where to look and what to look for

Many of the beetle and moth pests are small and their eggs and young larvae are very difficult to see. Even so, visual inspection of stores and artefacts is the most important part of any pest control strategy. It is vital to know how to look, where to look and what to look for.

All material on display or in storage should be inspected regularly. As the biggest resource involved is labour, this may require a compromise to concentrate inspection on certain areas of highest priority in the event of staff shortages. The priorities are material at greatest risk and of greatest value (historical or fiscal). The building should also be inspected and high risk areas identified.

Rather than a 'blitz' inspection once every few years, it is better to have a rolling system in which the building and its contents are systematically examined in accordance with a regular timetable. The inspection sequence should be decided using a detailed plan of the building, including display areas, stores and unused areas. It is essential to record all inspection details on a record form or in a log book (or use a spreadsheet package such as Microsoft Excel).

Concentrate on dark corners and dead spaces, particularly in attics and basements. Search for areas of poor hygiene and look at debris, birds' nests and old 'abandoned' or damaged displays or objects. Look on windowsills and other catchment areas such as air-conditioning ducts and filters which may be the home of clothes moths and carpet beetles. Historic houses are particularly prone to cluster flies which hibernate in the autumn. The piles of dead flies provide an ideal food source for carpet beetles – many cases of damage caused by carpet beetles to historic carpets and curtains have been traced to dead flies in attics and window voids.

Look in display cases and store boxes containing material at risk, e.g. wool garments and animal skins. Look under objects and move them if possible. Look into pockets or behind seams of textiles and clothing. Look inside wooden objects and behind the polished veneer front of furniture as beetles will often concentrate their attack on the poorer backing timber.

Examine display cases and racking or storage systems – these may be infested in their own right. Dead spaces under display cases and behind objects are a frequent source of hidden infestation.

Examine work rooms, dark corners with debris and food spillage in eating areas. Other introduced objects such as ornamental pot plants can be a particularly troublesome source of stray insects. Live animals which are used in displays will need feeding and their food can be a source of insects. Biscuit beetles and moths may live in dried animal food, and larvae or crickets used to feed reptiles may escape and become established in the building. The dead bodies may then provide a source of food for pests such as carpet beetles. Kitchens, restaurants and staff rooms can be important sources of insects and should also be inspected regularly.

Live insects are tangible evidence of infestation, but other clues are needed to pinpoint problems before live insects become obvious. Dead insects or insect fragments can be important signs but may also be misleading if they are left over from a long-dead infestation. Pellets of frass, silk webbing or the cast skins of larvae are frequently the first signs of infestation and fresh evidence of this sort suggests that a more thorough examination of material is needed. Failure to find any live insects may indicate that the infestation is no longer active and that there is no need for concern. However, it may also be that infestation is dormant or active at such a low level that it remains undetected. For this reason, a follow-up inspection after a short period will be needed to check for any additional signs. Inspection for seasonal pests such as adult carpet beetles and furniture beetles is best carried out in late spring and early summer when they are active and flying.

Active infestation of woodboring beetles may be shown by the presence of new holes or piles of fresh dust and frass. If the object has not been moved recently, this is a warning that there is active infestation which needs to be followed up immediately. However, old frass from a long-dead infestation will often fall out of exit holes when objects are disturbed or moved. Some insect species produce frass which is a distinctive shape. Frass from three types of woodborer beetles and drywood termites is shown in Figure 33.

Keeping objects and their environment clean and free from dust will make cast larval skins and fresh boring activity obvious – a dirty store will make it difficult to distinguish old infestation from new.

Because of the difficulty of detecting infestation on material coming into museums, many operate a quarantine/incubation system to identify infested material by breeding out insects if they are present.

In the gloomy conditions which frequently prevail in old houses, museums and stores, it is essential to use a bright flashlight for inspection. A methodical search is preferable to a sporadic examination, and as previously mentioned, results should be recorded. Carefully examine each object and make a note of any changes since the last inspection was carried out. When damage to valuable artefacts by fabric pests or woodborers is detected, a sketch or photographic record should be made to show whether, on subsequent examination, any additional damage has occurred. Because they are so small and secretive, it is still possible to miss insects, even in a thorough inspection. It is also logistically impossible to inspect all objects and areas in large buildings or stores – for these reasons insect traps should always be used to supplement visual inspection (Child 1999).

Insect traps

The development of insect traps for use in the public health and food industries has benefited museums because the traps have been successfully adapted for the detection of museum pests. The use of insect traps for monitoring is the first priority in setting up an IPM programme in historic houses or museums. Traps enable insects to be detected at very early stages of infestation and greatly improve chances of preventing damage. Over a period of time, a detailed picture of infestation problems can be built up to enable appropriate action to be taken (Child and Pinniger 1994; Pinniger *et al.* 1998). They can also provide a useful check on the effectiveness of control measures.

Sticky blunder traps

Non-specific sticky traps should be used for basic monitoring of insects such as silverfish, booklice, carpet beetles and other crawling insects. Sticky blunder traps are good for monitoring the residual population of insects in a building to assess the risk to the collection. They can also give a useful indication of the number of crawling insects which may be invading a room or building.

There are many different traps available, some of which perform better than others. Large traps designed to catch cockroaches are very successful for large insects such as oriental cockroaches and crickets, but because they usually have a folded and raised lip, they not very effective for small insects which have to climb over this to get in the trap. The simple, triangular, card types without a folded lip are preferable, some of which have a sticky patch under the trap designed to enable the trap to be stuck to the surface. This sticky area will often trap very small insects such as booklice or newly hatched carpet beetle larvae which cannot easily climb onto the top surface of the trap. The card traps are liable to physical damage and are ineffective when they get damp. In these situations a plastic trap with sticky insert can be used, but these low-height traps are less effective for catching clothes moths. Home-made traps made from sticky tape are rarely as effective for trapping insects as commercial traps.

Traps should be placed in a regular grid pattern against wall floor angles, preferably in corners where they will catch more wandering insects; traps which are low in height can be tucked under shelves and storage units. They should be checked regularly at least four times a year. The insects caught should be identified and recorded using a trap record form to simplify recording and analysing the data. Traps should be replaced when they become dirty, the adhesive has failed or if large numbers of insects have been caught – the trapped insects may provide food for pests. The more traps used, the more chance of finding pests, but a trapping programme should not exceed the resources available to check the traps. Match trap numbers to the priorities of the collection.

Pheromone traps

Many insects produce attractant chemicals called sex pheromones to lure mates. Some of the pheromones of museum pests have been identified and a few are available as synthetic lures to use with traps. The most successful lure is that for webbing clothes moth; others are available for the biscuit beetle, cigarette beetle and furniture beetle. Because of the varied behavioural response to pheromones, placement of these traps is more complex than simple sticky traps. The pheromones will generally only attract the target species and may even repel other insects. The lures are also far more expensive than simple blunder traps so they are usually only used in areas where there is a high risk to valuable objects or where one of these pest species is suspected. As the lures will generally only attract flying males, it is essential to have an understanding of the pest biology and the performance of the pheromone.

Webbing clothes moths (*Tineola bisselliella*)

Lures containing the two pheromone components have been very successful. The flight activity of the males is stimulated by the pheromone and suspended traps provide a very useful early warning of infestation. When temperatures are high enough to promote flight, up to 20 times as many moths may be caught on traps with pheromone lures as on unbaited ones (Cox *et al.* 1996). Below 20°C, the males are reluctant to fly and the suspended traps are relatively ineffective. In these situations, the pheromone can be placed on simple sticky traps on the floor; traps with lures will catch more male moths. There is some evidence to suggest that the lures also have a low-level attraction to the case-bearing clothes moth *Tinea pellionella*.

Furniture beetles or woodworm (*Anobium punctatum*)

A synthetic pheromone known as Anobinone is available but the results of trapping with this lure have been varied (Pinniger and Child 1996a). The pheromone-baited traps are sometimes extremely effective in trapping beetles, even at very low levels of infestation. However, there has sometimes been a very poor response on traps in some areas which were known to be infested. The success of the lure seems to depend upon the design of the trap, temperature, RH and light levels. We need to know far more about the behavioural responses of beetles at low temperatures to be able to use these traps to accurately pinpoint infestation and target control treatments.

Biscuit beetles (*Stegobium paniceum*)

Stegobinone, the pheromone of this species, is reported to be identical to that of furniture beetle *Anobium punctatum* (White and Birch 1987). Fuji market a trap containing a lure which is an analogue rather than the authentic Stegobinone; traps with these lures will catch more adults when they are placed in warm, well-lit areas where the adults are flying freely (Child and Pinniger 1994). At lower temperatures, the lures seem to be ineffective.

As the pheromones of *Stegobium* and *Anobium* are identical, the *Anobium* pheromone lures are also effective against *Stegobium*. In trials in two small museums, more *Stegobium* adults were trapped with the Anobinone lure than on traps without the lure. In one museum, an expensive and unnecessary roof timber treatment was carried out because the biscuit beetles on the Anobid traps were incorrectly identified as woodworm *Anobium punctatum*.

Cigarette beetles (*Lasioderma serricorne*)

The pheromone for this species is known as Serricornin. A number of successful traps have been designed for the tobacco industry, some of which have been used in museums (Gilberg and Roach 1991). A combination of the pheromone lure and a food lure in the Serricotrap enables the trap to attract male and female beetles. As with biscuit beetles, these lures only work well when it is warm and the beetles are flying.

Carpet beetles (*Anthrenus* sp.)

The pheromone of the varied carpet beetle, *Anthrenus verbasci*, has been identified and is available in lures in the USA. There is a different pheromone for the Guernsey carpet beetle *Anthrenus sarnicus* which was tested in museums in the UK but is not available commercially (Ackery *et al.* 1999; Finnegan and Chambers 1993). The pheromone of both species was far more effective when the insects were also attracted to light from windows.

Ultraviolet (UV) traps

These suspended or wall-mounted traps are attractive to wasps, moths and flies but are not so effective for museum pests. A UV trap with sticky boards has been developed to monitor death watch beetle populations (Ridout 2001) (see also p. 64).

General principles of insect trapping

- Survey the site, prepare a plan and decide where to place traps.
- Place sticky traps in a regular grid pattern.
- Date-label traps and mark their position on a plan.
- Place traps on floors in corners and near walls, not in the middle of open areas.
- Check traps at regular intervals (minimum four times a year).
- Identify and record insects caught on traps.

- Record whether insects caught are larvae or adults.
- Replace traps when they are dirty or if the adhesive has failed.
- The more traps used, the greater the chance of finding insects.
- The workload should not be underestimated and trapping programmes should be designed to be manageable.
- Pheromone lures and traps are much more expensive than the small sticky blunder traps but are very good for accurate monitoring and early warning in sensitive areas.
- Large numbers of non-pest insects may be caught on traps if they are near an outside door or window; the traps should be replaced more frequently or the trapped insects will become food for pests.
- Over a period of time a record of catch will build up a picture of the distribution of insects.
- Additional traps should be placed in areas where pests need to be more accurately pinpointed.
- Traps should be used as a supplement to visual inspection and the information used to target preventive and remedial measures.
- Traps can be used to check on the effectiveness of control treatments.

Wooden objects – X-rays, ultrasonics and incubation

It is extremely difficult to identify hidden infestation of some woodboring beetle larvae which, as previously mentioned, may take years to develop before the adult emerges from the wood. Examination of holes and damage in wooden objects can be confusing because of the many different ways in which the damage may occur. Ethnographic objects can be found with tunnels in the timber made by longhorn beetles when the tree was standing. These tunnels are usually large and the carving and form of the objects often indicate that the hole existed before the object was made. Damage by bark beetles may also be seen as fan-shaped patterns resulting from the long-vacated larval tunnels between the wood and the bark. More difficult to diagnose are the smaller holes caused by past infestation of felled timber by ambrosia beetles which now are of no concern. There may also be additional damage by termites to add to the confusion.

We need to know if the holes are a sign of active infestation by one of the woodborers such as *Anobium* or *Lyctus*. Small objects can be X-rayed to show activity but the interpretation of X-ray photographs is a skilled technique. Large larvae can be seen in tunnels but smaller larvae may be overlooked because of the density of the wood and the lack of clear opaque structures in the larvae. It is usually very difficult to distinguish larvae in objects with large-section timbers and large objects may be impossible to X-ray. Therefore, incubation is usually the only way of really deciding whether a suspected infestation is really active.

Active infestation in wood can be detected by the oxygen which is used and the carbon dioxide which is produced by the respiring larvae (Koestler *et al.* 2000; Stusek *et al.* 2000). This can be done easily in laboratory conditions, but the limitations of the sensing equipment currently available make it far more difficult to produce a reliable method that can be used in situ. A number of studies have been carried out on the use of ultrasonic or sound sensors to detect woodborers in structural timber and while heavy and active infestations of woodboring beetles can be detected, low levels can easily be missed, particularly when temperatures are low and insects are inactive. These systems are more effective for detection of termites which have a much higher level of activity than beetle larvae. One company even employed trained sniffer dogs which they claimed would smell out dry rot and termites!

CHAPTER FOUR

Prevention of insect infestation

The need for regular control measures as a response to the discovery of insects is an admission of failure. Effective IPM will include a strategy for the safe display and storage of museum objects which places the emphasis on prevention rather than cure. At first sight, the effort needed to initiate and maintain a preventive strategy seems far greater than that required to carry out fumigation or insecticide treatments. However, the advantages and savings of prevention become clear when the long-term benefits are examined.

Effective prevention requires:

- An understanding of insects and their habits.
- Adoption of a disciplined regime appropriate to the museum or house and the objects in its care.
- Exclusion of pests.
- Environmental awareness.
- Regular inspection and monitoring.
- Effective quarantine procedures.

Exclusion

The pests which attack objects on display or in storage can enter the museum in many ways. Adult beetles, such as furniture beetles, carpet beetles and clothes moths can fly in from sources outside the building. Adults and larvae may crawl in from adjacent, infested areas or objects. It is therefore important to try to keep uninfested objects free of pests by making it as difficult as possible for the pests to get at them.

Methods of keeping insects out include:

- Fitting doors and windows with unobtrusive sealing barrier strips to prevent the entry of larger insects.
- Fitting bristle strips to doors and sash windows.
- Sealing windows that are never opened.
- Fitting windows and doors with fly mesh screening (where appearance is not a priority).
- Checking pot plants, cut flowers and firewood which can all carry insects into storage or working areas.

These measures will keep out non-pest insects such as wasps, cluster flies and lacewings thereby preventing their dead bodies becoming food for pests.

Plants and shrubs on or near walls should be kept well pruned as they can create pest harbourage and provide a route into buildings. There has to be compromise with many historic buildings where removal of creepers and other climbing plants would be sterile and aesthetically unacceptable. Birds should be excluded from buildings as birds' nests provide food and harbourage for insect pests (see Chapter 6).

Modern purpose-built storage should be designed to minimise pest access.

- Keep the number of doors and windows to an absolute minimum.
- Where doors and windows are present, fit them with tight-fitting draught excluders.
- If doors and windows have to be kept open for reasons of ventilation, they must be fitted with fine mesh screens.
- Keep flowering plants and shrubs well away from the walls.

In many buildings it is not possible to completely exclude insect pests but a great deal can be done to prevent pests that do get in from attacking objects.

Environment

The influence of temperature and RH on the prevention of infestation is crucial as insects need warmth and humidity to thrive. There is frequently a conflict in display areas between the comfort of the public and the requirements to keep the objects safe. However, in stores it is sometimes possible to maintain conditions which are unfavourable to insects to prevent them becoming established. The main environmental objectives for insect pest prevention in storage areas are:

- Keep the temperature low.
- Keep the RH low.
- Limit variations in temperature and RH.
- Keep the environment and the objects clean.

These conditions can be maintained by using sophisticated and expensive apparatus to control the storage environment. However, it is also possible to achieve considerable improvements at low cost by removing sources of undesirable heat and moisture.

Temperature and RH

Warm temperatures of 20°C and above will encourage insect breeding so cool conditions are recommended. Areas near heating pipes and in ducts are often far warmer than the rooms they are supplying. Direct sunlight can cause localised hot-spots even in cool areas and uneven temperatures may result in localised condensation.

Many insects, such as biscuit beetles, will survive at low RH but some species thrive when it is damper. Furniture beetle or woodworm has declined in recent years due to the increased use of central heating which has reduced average RH levels. It is usually only found infesting wood in damp basements, attics or in outbuildings and in objects which have been stored at ambient conditions. Silverfish will only breed rapidly and cause serious problems in conditions of above 70% RH. As already stated, booklice also need higher levels of RH than is normally found in libraries and archives. RH levels should be measured and monitored and sources of dampness – such as condensation, poor damp proofing or leaks from gutters or water pipes – must be checked and remedied.

Hygiene

Because insects are small they may find sufficient food in relatively small areas which may not be immediately obvious – cleaning is 99% of successful pest prevention in an IPM programme. Pests thrive in a dirty environment and debris and rubbish provide shelter as well as food.

The display and storage areas of museums and houses should be kept as clean as possible. Apart from providing food and harbourage for the pests, dirt and rubbish can create microclimates which may trap heat and moisture and encourage pest development. Dirt and debris may also hide pests and prevent them being detected until serious damage has occurred. Visitors shed kilograms of skin and hair every year and this provides food for pests unless it is regularly removed.

Poor hygiene can lead to many pest problems. Old rodent and birds' nests can support spider beetles and carpet beetles. In many historic houses, old birds' nest material and debris has been left in blocked-up chimneys, heating ducts and other dead spaces providing a source of food and harbourage for insect pests which have then infested carpets, furniture and curtains. Bird, rat and insect droppings can also provide food for pests leading to the development of damaging bacteria and moulds. Mould can encourage mould-feeding insects such as silverfish, plaster beetles and booklice.

Cluster flies, particularly *Pollenia rudis* (Fig. 34) are a problem and a great nuisance in many historic houses. Their larvae live in grassland and the adults come into houses in their thousands to hibernate in the autumn, invading attics, towers and upstairs rooms. Many of them die and the dead bodies provide an ideal source of food for carpet beetles and spider beetles. It is difficult to prevent the flies getting in as they are very persistent and will find small gaps and crevices. Sash windows seem particularly attractive and are difficult to proof. Flies also accumulate in window heating boxes where they may be difficult to remove. If the flies cannot be kept out they should be removed by vacuuming at frequent intervals. Smoke generators, dichlorvos slow release strips (see p. 71) and spraying with residual insecticides will kill many flies but it is still necessary to remove the bodies. UV fly traps will catch cluster flies and may be of use in some areas. However, the traps must be cleaned out regularly and a UV source may not be desirable near historic objects. There is some evidence that lavender and lavender oil is repellent to cluster flies.

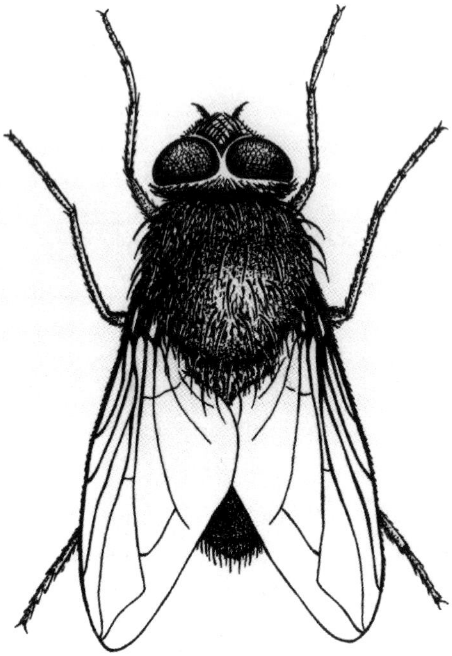

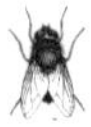

Actual size

Figure 34 Cluster fly *Pollenia rudis*

It is absolutely essential to remove and destroy unwanted material such as old felt-covered display boards that might act as a reservoir of infestation.

Areas to check and clean include:

- Unused chimneys and blocked fireplaces.
- Old bird, wasp and bee nests in attics.
- Old heating and ventilation ducts.
- Cavity walls and floors.
- Heater boxes.
- Unused rooms or cupboards, particularly in attics and basements.
- Gaps between walls and floors.
- Dead spaces behind and under storage cabinets, display cases and plinths.
- Dead spaces under and behind storage shelving.
- Felt lining on boxes and felt sealing strips on doors.
- Old and discarded display material, particularly when covered by wool felt.
- De-accessioned material which has not been removed.

It is unfortunate that cleaning and hygiene are usually given low priority compared with more glamorous projects. Many cleaning schedules are targeted on the most obvious public areas which may appear to be clean superficially. A close examination with a good flashlight will usually show accumulations of organic dirt and debris in corners, wall/floor angles and behind fittings which will support insect pests. Unused rooms and storage areas are often dirty and neglected. The investment of time and effort in creating a clean, pest-free environment will immediately benefit the care and conservation of objects.

Damage by micro-organisms and insects is greatly reduced on clean objects. Conservation of objects may require a compromise in the removal of dirt which may in itself have historic importance or may contribute to the strength of the article. Within these limits, however, objects should be as clean as possible. Gross dirt and debris should be removed from incoming specimens wherever possible before storage.

Storage

Objects in store can be protected from pest attack by storing them in ways that prevent the insects from getting at them. Insects will penetrate small cracks and crevices, but display and storage furniture can act as a barrier to pest attack if it is well designed and maintained. For example, insect-proof drawers for insect collections may prevent serious problems in the future. Some apparently well-fitting storage or display cabinets can lead to a false sense of security since small beetle larvae can gain access through minute gaps in joints and seals.

Cupboards, cabinets and drawers which may appear to be sound should be inspected because they may have hidden cracks which allow insect access. Some old cabinets and cases have seals made of wool felt providing food for clothes moths and carpet beetles – it should be replaced with a non-wool substitute or treated with a residual insecticide.

The simple expedient of enclosing an object in a clear plastic bag can keep many pests at bay, allow easy inspection of the contents. Additionally, if the object is subsequently found to be infested, the bag will keep the infestation from spreading to other objects. It is essential that, when objects are stored in sealed bags or containers, the conditions within the enclosure do not cause deterioration due to excessive moisture.

Any system of storage or display which gives some separation of objects from each other and the surrounding environment will reduce the chances of infestation. However, if objects are hidden away in containers, it is essential that high-risk items are inspected regularly.

The trend towards 'open' museum displays and dioramas carries increased risk of insect attack because of the vulnerability of objects and the impossibility of isolation and segregation. Increased vigilance and regular inspection of these displays is necessary. It is also essential to check materials used in the display such as logs, straw, dried vegetation, ersatz food, costume and mannequin hair.

Inspection

Details of specific inspection and monitoring techniques are given in Chapter 3. These general principles must be applied to both the building and objects on display and in storage. Items are generally far more at risk in storage than when on display. However, pests will attack displayed material when it is dark and undisturbed. Some examples of places which should be regularly checked are:

- Costume or curtains touching the floor.
- Heavy furniture placed on wool carpet.
- Rugs placed on wool carpet.
- Wool felt under objects.
- Natural history specimens in dioramas.
- Wool felt in keyboard instruments.

It is far easier to see signs of insect activity, such as frass, cast skins and dead adults, in display cases with plain light-coloured bases than in those with a granular or textured finish. All high-risk items (see p. 7 for a list of these) in store must be regularly inspected to keep a check on the success of the exclusion strategy – a daunting task especially when these are in boxes. Marking the boxes which contain the most vulnerable objects can help prioritise the inspection.

Many objects are composite and it may not be immediately obvious that they contain infestable material. Insects may attack adhesives such as animal glue and starch paste and cause incidental damage to the surrounding material. Modern adhesives and other synthetic materials are not nutritious and some may even be poisonous to the insects that try to eat them.

Quarantine

Material which is newly accessed, on loan or being loaned is probably the prime route for introduction of insect pests. In order to prevent pests being introduced, the following strategy should be adopted.

All organic material and objects coming into the museum must be placed in a designated area and checked for signs of infestation. If there is active infestation then objects should be treated as soon as possible.

If there is no direct evidence of live insects but there is reason to suspect that there may be active infestation, then objects should be isolated and incubated over a summer period. If adult insects are seen to emerge, then remedial action can be taken (see Chapter 5). The details of quarantine operation can be adapted to fit the types of objects and the needs of a museum. For example, some natural history museums treat all incoming plant and insect material by freezing before it is admitted to the collections.

Treatments which can be carried out on-site in the quarantine area by museum staff:

- Freezing.
- Ageless anoxia.

Treatments which may need a special area or be carried out off-site by specialist staff:

- Thermo Lignum heat treatment.
- Treatment with cylinder nitrogen or carbon dioxide.
- Fumigation with methyl bromide.

See Chapter 5 for details of these treatments.

CHAPTER FIVE

Control of insects

What happens when preventive measures fail and insects are found? A panic-stricken grab for the nearest can of insecticide is not the answer. A rational and carefully considered reaction is required to decide the best course of action.

- Are the insects pests or non-pests?
- Are the insects alive or dead?
- Are there many insects or only a few?
- Are the insects widespread or restricted to one area or object?
- Are there insects in the building?
- Are there other objects at risk in the area?
- Can the object be removed or the area isolated?

Effective control of insects can be achieved in a number of ways using physical or chemical methods. If it is decided that control measures are needed then it is important to choose the most appropriate course of action. Methods which may be effective for killing insects in the building and preventing their spread may be totally inappropriate for the treatment of infested objects.

Options for treatment

Insects in buildings

- UV traps
- Lower temperature
- Lower RH
- Residual insecticides
- Fumigation

Insects in objects

- Physical removal
- Freezing
- Heat
- Anoxia
- Carbon dioxide
- Fumigation
- Residual insecticides

Insect control in buildings

UV traps

Widely used in the food industry and public health to control flying insects, they can be used in museums and historic houses but they have serious limitations and can present risks. They are designed to be attractive to wasps, houseflies and blow flies and when properly sited and maintained can be very effective for these species. They are not particularly attractive to many museum pests although some may get caught on the traps. Their main value is to catch flying insects in catering areas, although they have also been used to catch cluster flies in historic houses and stores. There is evidence that female death watch beetles are attracted to UV light and UV traps with sticky boards have been used to monitor pest populations of this species (Ridout 2001). Most UV traps have catch trays which fill up with dead flies, wasps and large moths – if they are not regularly emptied and properly maintained UV traps can cause more problems than they solve. In addition, the UV source may not be acceptable near historic objects and the types with electric killing grids may also present a fire hazard in some areas.

Environment – temperature and RH

The environment can be manipulated to control insects in buildings but the tolerance of many insects may limit the effectiveness.

Although temperatures well below freezing are required to kill most pest insects, temperatures below 5°C will effectively prevent insects from developing or breeding. Temperatures between 5°C and 15°C may allow a few species to develop very slowly but most pests will not increase in numbers. It therefore follows that if objects are stored at these temperatures, serious damage will be prevented. The principle of keeping high-risk valuables such as fur coats in cold stores to prevent damage by beetles and moths has been practised for a long time but it is only relatively recently that the same procedure has been adopted for objects in museum stores. Although effective, specially constructed cold stores are expensive to build and run and may create associated problems with high RH. A great deal can be done to improve the general storage environment by conditioning air to lower temperatures, but there must be very careful monitoring of local RH and temperatures to avoid creating more problems. Where resources or practicality do not allow full scale air-conditioning, risk of pest attack can be minimised by eliminating local high temperature areas i.e. avoid positioning objects in direct sunlight or near central heating pipes and radiators.

RH and its control is an extremely important aspect of museum display and storage. The level of RH of a closed volume of air is directly related to its temperature but as stated previously, many of the pest species can tolerate a wide range of RH. Extremely dry conditions of less than 10% RH are unfavourable for most insects but this is unlikely to be achievable or acceptable for stored objects at risk from insect attack. However, keeping the RH levels to below 55–60% will prevent furniture beetle breeding. Simply recording RH in the centre of a room may be misleading as local microclimates with higher water content than the general environment may be present. Spot-checking near floors and north walls may show damper areas which will allow insects to survive. Damp conditions, in excess of 75% RH, encourage mould-feeding insects and mites and should be avoided.

Biological control

One of the most frequently asked questions is, 'can we use biological control for pests in museums?' Unfortunately, the answer has to be for all practical purposes – no. The reasons are many, but are predominantly governed by the normal requirement for low or zero tolerance of pests in museum environments. Most classic biological control systems using predators or parasites are based on a balanced relationship. This demands the presence of a low level of prey to ensure the continuity of the system. Sometimes the pest has to be released in low numbers with the predator or parasite to maintain the equilibrium; generally this would be unacceptable in collections areas. There is also very little information on parasites and predators of most museum pests. Some work carried out on *Lalius*, a very small parasite of Dermestid beetles, showed that each female wasp could parasitise about 15 *Anthrenus* larvae. A proposal to mass-rear and release these tiny wasps in a museum was rejected on grounds of cost and perceived reaction of staff to the release of wasps. Some predators are very efficient, but only at high densities of pests. For example, numbers of small metallic blue-green *Korynetes* beetles near woodworm holes are a sure sign of a heavy *Anobium* infestation.

BT, the active toxicant from *Bacillus thuringiensis,* will kill clothes moth larvae but this is the equivalent of a residual pesticide rather than a biological system. There is the possibility in the future that some biological systems based on fungal or bacterial pathogens may be suitable for museum use: the limiting factor seems to be that the effective pathogens only work really well at high temperatures and RH, the very conditions not encouraged in collections. Subterranean termites live in a more pathogen-friendly environment and this may be a more encouraging area for biological control. A nematode pathogen has already been developed for termites which has achieved some success.

Chemical control methods

The use of chemicals to protect food and objects dates back to Greek and pre-Roman times when various plant and mineral extracts were used as insecticides. In some countries many traditional techniques based on natural materials, such as neem, are still in use. The general basis of all these methods, whether based on synthetic or naturally derived materials, is that they are poisons. Not only are they poisonous to insects, but they are nearly always poisonous to other animals including man. A balance must be maintained between the toxicity of pesticides to the target insects and the risk to man, objects and the environment.

Rational use of pesticides as part of a control strategy offers the means to eliminate insect pest problems from museum environments (Pinniger and Child 1996b). Chemicals should be targeted to control an outbreak of an existing and known infestation or, as a prophylactic treatment to provide a toxic environment that will kill any insect which presents a possible hazard. Indiscriminate and unnecessary use of pesticides may jeopardise the future of the collection, the health of the staff and also create environmental hazards.

Large quantities of liquid insecticides applied to structural timber to kill woodborers were frequently solvent-based to aid penetration into the wood. These treatments kill emerging beetle adults and any larvae and pupae near the surface of the wood. However, they are unlikely to kill extensive and deep-seated infestation in wood. Treatments against woodborers should always be targeted on areas of active infestation as it is pointless applying pesticide to wood where the infestation is long-dead.

There are a bewildering number of insecticides in cans, bottles and packets. Not only do they contain many different chemicals, but there is also a wide variety of methods of applying them.

Space treatment

The insecticide is dispersed into the air to treat an infested area where it contacts and kills flying or crawling insects. There is very little persistence of the killing effect and most are inappropriate for use in museums or houses because they will leave a chemical deposit on all exposed surfaces.

Smoke generators

These were widely used for the treatment of inaccessible places such as roof voids which could not otherwise be treated. Usually containing permethrin, they are effective in killing cluster flies and have also been used against woodborers such as death watch beetle. Smoke generators need no application equipment and are relatively cheap. However, precautions are essential as they are pyrotechnic and present a significant fire risk. It is always advisable to inform the local fire brigade because, in addition to the fire risk, clouds of smoke may be seen escaping from the roof of the building which may prompt a 999 call.

Mists and fogs

These must be applied with special equipment and should only be used for controlling flying insects such as flies and mosquitoes.

Vapour

Slow-release strips which emit dichlorvos (see p. 71) over a period of time are very effective for the treatment of enclosed and unventilated areas such as display cases. They should not be used in open working or public areas.

Baits

Some very specific insecticides are formulated as baits so that the insect dies when it consumes the insecticide in the bait. Examples include hydromethylnon for cockroaches, and fipronil for termites (Quarles 1997) (see Table 3).

Residual treatments

The insecticide is applied to surfaces of the building or to display and storage furniture. The insecticide kills insects at the time of treatment but the main objective is to leave a residual deposit on surfaces which will remain lethal to adult or larval insects that crawl over it after a period of some weeks or even months (Morgan *et al.* 1993).

Liquid applications

Some are ready-to-use liquids but most general-purpose insecticides for structural treatments are formulated as wettable powders or emulsifiable concentrates so that they can be sprayed after dilution with water. There are also some specialised gel and mayonnaise-type formulations for treatment of structural timber.

Dusts

Residual powders for application to dead spaces, cracks and crevices.

Chemical names of insecticides

Before an insecticide is used, the effects of the chemical on insects, objects and staff must be considered. For this reason it is important to know the chemical groupings of insecticides and their general properties. In addition, some of the solvents or dilutions used in some formulations can cause staining or damage to objects.

Much confusion is caused by the multiplicity of trade or brand names given to insecticides including such gems as Bug-off, Croakaroach and Zu-i-cide. The important fact which must be displayed on all pesticides in many countries, including the UK, is the standard name of the insecticide or insecticides which are contained in the formulation. Whatever the brand name on the can, it is important to know the name of the active ingredient(s). Therefore, each chemical will have one chemical name which describes the structure, one accepted standard name and maybe dozens of different brand names. For example the chemical 3-phenoxybenzyl-3-(2,2-dichlorovinyl)-2, 2-dimethyl cyclopropanecarboxylate is a synthetic pyrethroid with the standard name of permethrin. There are many brand names for permethrin including Constrain, Coopex, Ambush and Muscatrol, all registered for different uses.

Pesticides must only be used as directed on the label and under the UK Pesticide Regulations, it is now an offence to use a pesticide in a way which is not registered or approved.

The major insecticide groups and the advantages and disadvantages of some examples are given in Table 2. A few of the more commonly used insecticides will be described in more detail.

Natural pyrethrins/synthetic pyrethroids

This is a much misunderstood group of chemicals. Natural pyrethrins have been used for many years and are obtained from the flowers of a species of chrysanthemum. They produce a rapid knockdown and kill of insects and, because they are degraded very quickly by UV light, they do not leave hazardous residues. However, the synthetic pyrethroids which have been designed and manufactured to simulate the action of pyrethrins fall into two distinct categories.

Table 2 Insecticides*

Group	Examples	Properties
Organo-chlorines (OCs)	DDT Dieldrin Chlordane Gamma HCH (lindane)	Residual insecticides which are extremely persistent in the environment. Use restricted or banned in many countries.
Organo-phosphorus (OPs)	Malathion Fenitrothion Chlorpyriphos	Used for residual treatments. Can be persistent on surfaces. Most of those now in use have a relatively low toxicity to man.
	Dichlorvos (DDVP, Vapona)	Used for space treatment. Restricted in some countries because of concerns over safety.
Carbamate	Bendiocarb Propoxur Carbaryl	Used for residual treatments. Can be persistent on surfaces.
Pyrethrins and synthetic pyrethroids	*Non-residual* Natural pyrethrins Bioresmethrin Phenothrin Tetramethrin	Used for space treatments. Non-persistent, broken down by UV light and do not cause residue problems.
	Residual Permethrin Cypermethrin Alphacypermethrin Cyphenothrin Cyhalothrin	Used for residual treatments, persistent on surfaces. Wide range of toxicity to man but the most potent are used at very low dosages.
Juvenoid insect growth regulators (IGRs)	Methoprene Hydroprene Fenoxycarb Pyriproxyfen	Not poisons, they disrupt the development cycle of insects. Very low toxicity to man.
Chitin synthesis inhibitors (IGRs)	Hexaflumuron Flufenoxeron Diflubenzuron	Not poisons, they disrupt the development of the insect cuticle. Generally low toxicity to man.
Miscellaneous	Boric acid	Residual dust or a bait poison.
	Hydramethylnon	Slow-acting but very potent poison bait.
	Naphthalene Paradichlorobenzene Camphor	Repellents which are only toxic to pests at very high doses. Use restricted in some countries because of safety concerns.
	Desiccant dust	Residual dust with a physical action.

* NB Not all of these insecticides are approved for use in museums in the UK

Those which are degraded rapidly in the same way as natural pyrethrins are called the **non-persistent pyrethroids**. Many of these, including bioresmethrin and phenothrin, are less toxic to man than natural pyrethrins. When formulated in a combination which produces the maximum effect of knock-down and kill they are extremely effective for controlling flying insects.

The other group, called the **persistent pyrethroids**, have been designed to give rapid knockdown properties together with a residual life on surfaces. This has been achieved at a cost of increased toxicity to man and although their extremely high toxicity to insects means that they may be used at very low doses, they are do not fall into in the same safety category as natural pyrethrins. Additionally, some persistent synthetic pyrethroids, such as cypermethrin, may cause allergic reactions. Although relatively costly compared to some other chemicals, they are potent insecticides. Permethrin and cypermethrin in particular, have been shown to be extremely effective against many insect pests, including woodboring beetles.

Gamma HCH (Lindane)

Gamma HCH is an organochlorine insecticide which has been in worldwide use since the 1950s. It was an extremely effective insecticide for surface or injection treatment of wood and this, together with its relatively low cost, resulted in its widespread use. The concern over the persistence of organochlorine insecticides, including DDT and gamma HCH, has led to a steady decline in the use of this insecticide. It is a carcinogen that is now banned in many countries.

Dichlorvos (DDVP, Vapona)

Although a member of the organophosphorus group, dichlorvos has very different properties and uses. It vaporises very readily and although it will not penetrate deeply, it will pervade air spaces. Insects are killed when there is sufficient concentration of dichlorvos in the air. It is commonly used in the form of slow-release strips for treatment of rooms and display cases where the strips will give control and protection for up to six months. Dichlorvos can be extremely effective when used properly but the vapour must be allowed to build up to a level which is lethal to insects. The strips will not work if there is rapid air change, open doors and windows or air-conditioning. There is concern over exposure of staff to dichlorvos vapour and the deleterious effect of the vapour on artefacts at high RH and concentrations. Its use has now been restricted in the UK – and even banned in some countries – and it is essential that dichlorvos is used with proper regard to the registration regulations and recommendations for use on the label.

Insect repellents

Paradichlorobenzene, naphthalene and camphor have a long history of museum use. The solid material gradually vaporises and although the resultant vapour is repellent to some insects it is toxic only at very high concentrations. Naphthalene or paradichlorobenzene can give protection against attack by clothes moths but some of the beetle pests are much more tolerant, particularly if the concentration of vapour is allowed to fall (Linnie and Keatinge 2000). Contact with these materials may cause damage to resins, rubber and varnish or paint. The long-term toxicity to man of both naphthalene and paradichlorobenzene has given cause for concern in many countries where their use is now restricted or banned. High levels of paradichlorobenzene and naphthalene are still present in some museums from earlier treatments to protect collections.

Insects are repelled by high doses of some traditional plant-derived materials such as tansy, lavender and cedarwood oil. Some are currently being re-examined for efficacy and repellency and may have a valuable role in the future. It must be remembered that some botanical chemicals are as toxic as synthetic pesticides and the oils may have deleterious effects on objects.

Insect growth regulators (IGRs)

This term is used for insecticides which are not contact poisons but affect the development of the insect. One mode of action, called 'juvenile hormone analogues', acts through the insect's hormone system. These disrupt the development process and usually result in death of juveniles or sterile adult stages. They include methoprene, hydroprene, pyriproxyfen and fenoxycarb (Edwards 1993). The other main group, called 'chitin synthesis inhibitors', disrupt the deposition of the insect's exoskeleton so that the insects die when they moult. Examples include flufenoxeron and diflubenzuron. Although IGRs have great potency against many pests, there has been little published work of their efficacy against museum pests except for some trials of fenoxycarb in Australia. Hexaflumuron is registered as Sentricon™ in the USA and other countries for the control of subterranean termites. The Sentricon™ process involves bait monitoring for termite activity and then baiting with hexaflumuron at active sites. The process can be very effective but requires careful monitoring and management (Su *et al.* 1998). ***None of these insecticides are currently registered for use against museum pests in the UK.***

Desiccant dusts

These dusts are fine particles of diatomaceous earth or silica which damage the surface layers of insect wax and body cuticle and cause the insect to lose water and die of desiccation. The dusts are more effective at low RH and will remain active for long periods of time. As they are not poisons, there is little hazard from residues and they are therefore very useful for treating under cabinets and shelving in wall cavities, basements, attics and other dead spaces. Some formulations also include natural pyrethrins to give added rapid knockdown of insects.

Fumigant gases

Gases such as methyl bromide, sulfuryl fluoride and phosphine have been used to fumigate entire buildings. This is done by sealing the whole building or covering the building with gas-proof sheets. It can be very effective but is expensive and disruptive (the problems and limitations of fumigant gases are discussed in more detail on p. 78).

Insect control in objects

Non-chemical methods

Physical control

Destruction of objects is not a viable option for objects entrusted to the care of the museum. But if infestation is restricted to non-accessioned material, display materials, storage furniture or the museum structure, then it may be cheaper and more desirable to destroy the affected material and replace it rather than treat it with insecticides.

Careful inspection and removal of insect larvae, pupae and adults from objects may be a viable and effective method of control. The limitations depend upon the pest species involved, the nature of the object and the ease with which it can be inspected. In some cases vacuum cleaning can be used to supplement the inspection. If the technique is applied to materials such as textiles, the examination and cleaning must be repeated to ensure that any larvae hatched from eggs which escaped the first inspection are removed. Most textile dry-cleaning processes using organic solvents will kill insect pests. Objects should not be cleaned without seeking the advice of a conservator to ensure that the integrity of the object is not compromised.

Freezing

Freezing kills insects by rapid temperature change (Strang 1992). It is a widely used treatment for many objects, such as natural history specimens and textiles, and when proper procedures are observed, there is no damage to objects (Nesheim 1984; Miller and Rajer 1994; Petersson 1998). However, composite, fragile or unstable materials should not be subjected to extremes of temperature (Florian 1986, 1990).

Temperatures in the freezer should be –18°C or below. It is quicker to use a deep-freeze unit capable of reaching –30°C rather than a common household freezer – self-defrosting freezers should never be used (Gilberg and Brokerhof 1991; Strang 1996). If the items are stored in a cold climate and then frozen very slowly to –18°C, some insects may become acclimated and will not succumb to freezing. Therefore, the freezer should be capable of reducing the temperature in the objects within 24 hours. Even with a powerful freezer it takes longer for the cold to penetrate large objects or tightly rolled tapestries and skins. It is not sufficient to rely on the thermometer on the freezer as this may not reflect the conditions in the material being frozen. The temperatures in the centre of test materials should be recorded to check that the target temperature has been achieved. It is not advisable to freeze wet specimens unless this is to prevent bacterial and fungal decay.

The objects must be sealed in plastic bags to prevent damage by humidity change and moisture migration. Better results are obtained if the air volume around an object is as small as possible and so it is helpful to remove air from

Table 3 Options for treatment of objects

Treatment	Advantages	Disadvantages
Fumigation with toxic gas		
Methyl bromide *Sulfuryl fluoride*	Rapid treatment (24–48 hrs).	Methyl bromide banned in some countries. Residues in objects, may cause taint or corrosion. High toxic hazard.
Phosphine	Treatment 5–14 days. No application equipment.	Not effective at low temperatures. May corrode metallic surfaces. High toxic hazard.
Fumigation with carbon dioxide	Much lower hazard than toxic gases, very safe for objects. Useful for large objects under sheets.	Long (3 week) exposure at lower temperatures. Large bulk of CO_2 needed for large objects.
Anoxia (low oxygen)		
Cylinder nitrogen	Completely inert therefore very safe for objects and staff.	Long (2–3 week) exposure. May be ineffective at low temperatures. Difficult to maintain low oxygen levels with large objects which also need many bulky cylinders.
Oxygen scavengers	Very convenient for smaller objects. Safe for objects and staff.	Long (2–3 week) exposure. May be ineffective at low temperatures. Expensive for larger objects because of the large number of sachets needed.
Nitrogen generator	Very convenient for smaller objects. Safe for objects and staff.	High capital cost of equipment. Slow output of nitrogen.
Freezing	Quick (3 days) at –30°C, 7 days at –25°C, 14 days at –18°C. Safe for many objects and staff.	Objects must be bagged. Not suitable for objects which are very fragile or under stress.
High temperature	Very rapid, 24 hours cycle at 52°C. Safe for many objects and staff.	Needs control of humidity. Not suitable for objects which are very fragile or under stress.

the bags before they are sealed. An absorbent organic buffering material such as paper towels can be added to the bag to help stabilise the humidity. Place the object in the freezer for at least three days at –30°C or seven days at –25°C. In the past it was thought a double freezing at –18°C was necessary to kill all stages of insects but more recent work has shown that a single exposure for at least two weeks at –18°C will kill all of the pest species.

After the treatment, objects must not be unsealed until they have reached room temperature to prevent condensation on the object. However, if there are many insects in the building then objects can be left in the polythene outer to provide some protection from further insect attack.

Meticulous management is needed for the operation of a freezing regime in order to achieve the best results and prevent damage to objects. For many museums and historic houses, freezing is one of the most widely used techniques to prevent pest damage successfully without pesticide contamination (Crisafulli 1980; Berkouwer 1994).

Heating

Heating will kill insects much more rapidly than freezing but it is essential to ensure that elevated temperatures do not harm objects. Many insect pests will survive temperatures up to 40°C and unless great care is taken, the effects of higher temperatures needed to kill insects may be more damaging to objects than the pests. Most pests are killed in a few hours at temperatures of 55°C and above and washing textiles in water above 60°C will usually kill developing insects (Strang 1992; Hagn and Vagn Jensen 1996, 1998).

In the past, a number of museums used ovens to disinfest insect collections and this often resulted in brittle specimens and cracked drawers. Recent work has shown that damage due to shrinkage and distortion can be eliminated by controlling the RH around the object. A commercial Thermo Lignum process has been developed and is now used in Europe for treating objects (Child 1988, 1994; Pinniger 1996). This uses a special chamber in which the RH is accurately maintained at a set level during the entire 18-hour cycle from ambient up to 52°C and back. Many objects including books, furniture, textiles and natural history specimens have been treated successfully without damage.

Work is currently in progress to develop and evaluate heating techniques using bags to contain objects. The principles of RH stability are similar to those when freezing an object and robust objects have been treated successfully without problems (Brokerhof 1999; Xavier-Rowe *et al.* 2000; Strang 2001). The use of solar heat for treating objects in bags offers some exciting possibilities for developing countries, but more trials are needed before it can be recommended as a general treatment. As with low temperature treatments, it is inadvisable to subject composite, fragile or unstable materials to extremes of high temperature.

Microwaves have been used for rapid treatment of books, papers and herbarium specimens but there can be undesirable side effects as the heating may be uneven and localised overheating may occur. In addition, unnoticed metallic objects such as paperclips may cause sparking and ignition of specimens and paper (Philbrick 1984).

Atmospheric gases – modified atmosphere treatments

Conditioning of the storage atmosphere by the use of naturally produced carbon dioxide has been used for grain storage for centuries. Insects require oxygen and produce carbon dioxide; manipulation of the levels of these two gases can kill insects and prevent damage (Valentin 1993; Kigawa *et al.* 2001). Modified or controlled atmosphere treatments (CAT) have been developed as a direct replacement for fumigation of objects with toxic gases such as methyl bromide and ethylene oxide. The procedures used are in some cases very similar but special techniques have to be used as both carbon dioxide and nitrogen have much smaller molecules than the toxic gases (Selwitz and Maekawa 1998).

Nitrogen

Nitrogen has proven to be a very effective and safe method for treatment of sensitive objects (Gilberg 1989; Hanlon *et al.* 1992). The procedure kills insects by anoxia – the exclusion of oxygen – and therefore oxygen levels must be very low. Ideally this should be less than 0.1% but in practice 0.3% can be effective (Gilberg 1991; Rust *et al.* 1996; Bergh 1998). These low oxygen levels can only be achieved in a special chamber, or in individual bags manufactured from a specially made oxygen barrier film. The speed of the treatment in killing insects is dependent upon the pest species and the temperature. At 25°C and above, an exposure of two to three weeks should be sufficient to kill all stages and species of pests. However, at temperatures of 20°C, long exposures of four or five weeks may be needed to kill woodborers such as the furniture beetle, *Anobium punctatum*, which are naturally tolerant to low oxygen and high levels of CO_2. Anoxic treatments are not usually recommended at temperatures lower than 20°C because of the length of exposure needed.

- Large objects can be treated using nitrogen from cylinders or a generator.
- Small objects can be treated in sealed bags using an oxygen scavenger such as Ageless™.

Building a nitrogen treatment chamber may be expensive because of the need for absolute gas-tightness (Daniels 1997). Conversion of an existing fumigation chamber used for ethylene oxide or methyl bromide is feasible but can also be expensive because of the need for additional sealing and pipework. The nitrogen gas cylinder and bag method is cheaper to set up initially although an accurate oxygen meter must be purchased (Elert and Maekawa 1997). There are many different types of oxygen barrier film available, such as Nylar®, Aclar® and Marvelseal®, with different properties. All are sufficiently gas-tight for disinfestation treatments but those with the lowest oxygen diffusion rates are more suitable for long-term anoxic storage. Bags should be purpose-made for each treatment using a heat sealer. Low levels of oxygen can be reached quite quickly in small bags but larger enclosures will require repeated flushing with gas. Sucking residual air out of the bag before each flush with a vacuum cleaner can be much quicker and more economical than a continual flow.

The RH of nitrogen gas is very low and at levels of 99.9% would dehydrate the materials being treated. It is therefore essential to add a simple humidification system to the nitrogen gas supply line – the 'Getty' three-bottle system is the simplest and easiest to use (Selwitz and Maekawa 1998). Gas cylinders must be used and stored in compliance with local fire and safety regulations. Some large-scale treatments have been carried out using liquid nitrogen but the containers are bulky and the nitrogen must be passed through a heat exchanger.

Nitrogen can also be produced from a nitrogen generator using a compressor-driven oxygen filtration or absorption system. Very safe to use, they avoid the need for handling large and heavy gas cylinders. However, nitrogen generators can be complex and expensive and the output rate at low oxygen levels is low. Their big advantage is that when large quantities of gas are being used to flush large containers, the outlet gas is recycled through the generator. This method therefore releases small quantities of extra oxygen into the workspace rather than large volumes of nitrogen.

Oxygen scavengers

A number of museums now use a gaseous oxygen scavenger for treatment of individual specimens. Ageless Z™, the best known of these, is composed of moist active iron oxide powder encased in a porous packet. Oxygen in the atmosphere penetrates the packet and further oxidises the powder. A slight amount of heat and moisture is produced by the reaction, but if the packets are spaced apart and kept out of contact with objects then it does not build up enough to damage the material being treated. The small amount of additional moisture will have no measurable effect on well-buffered absorbent materials such as textiles or natural history specimens. More sensitive non-absorbent material can be wrapped in buffering cloth or acid-free tissue paper.

Ageless™ will work effectively only when the object to be treated is enclosed in a bag made from oxygen barrier film. It is useless to use ordinary polyethylene as this is porous to oxygen. The bag should be big enough to accommodate the object to be treated with a few inches excess which will become the seam of the bag when it is heat-sealed. The number of Ageless™ packets to be used should be calculated from the type of packet and the volume of the bag. An indicator called 'Ageless eye' can be used to show that levels of below 0.1% oxygen have been achieved (Gilberg 1990). Objects should be left for at least three weeks at 25°C to kill all stages of insect pests. With this system, set-up costs are lower than with the cylinder treatment but costs of Ageless™ may be high if large objects are treated. A combination of nitrogen cylinder flushing and Ageless™ can be used for the treatment of large objects (Newton *et al.* 1996). This is a practical way of ensuring that slow leakage of oxygen into the bag is absorbed by the Ageless™ and does not allow insects to survive.

Ageless™ will also slow any degradative processes requiring oxygen, such as mould growth or oxidative chemical reactions. Other types of Ageless™, such as RPA, will absorb oxygen, corrosive gases and moisture and therefore have applications for long-term storage and collection care. These special types are far more expensive than Ageless Z™ and it is normally not economical to use them for insect control in large objects.

Carbon dioxide (CO_2)

Carbon dioxide treatment has been accepted in the food industry for many years and the technique has been adapted for use by museums (Newton 1993). The treatment procedure used for CO_2 is similar to that used for nitrogen in chambers or bubbles. As with nitrogen, the gas works slowly, particularly at low temperatures, and some treatments add slight heating of the chamber to 30°C to increase insect metabolic rates. Levels of CO_2 need to be about 60% and this requires the use of a meter to monitor the percentage of CO_2 in the chamber during the exposure period. There is evidence that CO_2 is more effective in killing insects when oxygen is present and so it is counter-productive to use levels above 90%. Unlike nitrogen, carbon dioxide is poisonous and therefore small leakages of oxygen into the enclosure will not affect the treatment, making CO_2 far more practical than nitrogen for treatment of large objects and enclosures (Binker 2001). However, it is important to check on local safety procedures and regulations before proceeding as there may be some restrictions on access to treatment areas. Leakage of carbon dioxide into the treatment area can be dangerous and must be monitored with a CO_2 detector. In some countries, including the UK, CO_2 is registered as an insecticide and can only be used by professional operators. There have been some concerns that CO_2 will react with water to produce carbonic acid, but there is no evidence that this will happen at the usual range of RH and moisture contents used for treatment of museum objects.

Other gases

Argon and other inert gases have been used with varying degrees of success but because they are more expensive than nitrogen or carbon dioxide, it is difficult to justify their use (Valentin 1993).

Radiation

Many forms of radiation have been investigated for purposes of pest control but very few techniques have progressed beyond the purely experimental. Microwave radiation is in common use in domestic microwave ovens and some work has been carried out to investigate the effects of subjecting foodstuffs to microwave treatment. The results indicate that the major effect is the rapid heating of the food which kills the insects. Insects alone can survive quite long periods of intensive microwave radiation and it is doubtful whether the technique could be applied to many museum objects because of the risks of damage caused by localised heating.

Gamma radiation penetrates objects and is becoming increasingly used for sterilising food and medical items. As gamma radiation leaves no residue and does not make irradiated materials radioactive, it could find use as a replacement for fumigation (Horokova and Martinek 1984; Hanus 1985). It has also been used to control decay in waterlogged timber. However, there are many drawbacks, including the high cost associated with a gamma radiation unit and the safety restrictions and concerns in some countries which would not allow it to be used. There is also evidence that gamma radiation may adversely affect plastics and glass. It is likely that the doses of gamma radiation high enough to kill insect pests would degrade museum objects.

Residual insecticides

The practice of treating infested objects with liquid insecticides to kill woodborer infestation was formerly widespread. Insecticides such as gamma-HCH (Lindane) and permethrin were applied to surfaces or injected into flight holes; these were frequently solvent-based to aid penetration into the wood. These treatments kill emerging furniture beetle adults and larvae and pupae near the surface of the wood. Such treatments may be appropriate for bulky and robust objects but are far less suitable for delicate objects and fine furniture. Insecticides cause deliberate contamination of the objects and some solvent-based products also cause damage to varnish and other wood finishes. The recent development of permethrin micro-emulsion formulations enables objects to be treated with far less risk of solvent damage. Where it is not practical to treat objects with alternative non-chemical methods, then spraying with an appropriate and safe residual insecticide will give some degree of control. However, residual insecticide treatments are unlikely to kill extensive and deep-seated infestation in wood. Treatments should always be targeted on areas of active infestation as it is pointless applying pesticide to wood where the infestation is long-dead.

Fumigation

This is a term which is frequently used for almost any insecticide treatment but which should only be applied to the use of fumigant gases. Objects may be fumigated under gas-proof sheets or in specially built or modified chambers. In some cases whole buildings and their contents can be fumigated by sealing all doors, windows and other apertures with special tape or by covering the entire building with sheets. The development of a fumigation 'bubble' which is portable and has a specially designed inlet and venting system has been widely used in museums.

The principle of fumigation is that the gas is introduced into the air space around objects or in a store and must be contained for a period of time in order for the gas to penetrate the material being fumigated and to poison the insects. When carried out correctly, fumigation can be guaranteed to kill 100% of the insects even when they are deep-seated in objects. After exposure, the gas must be aired off and, because the gas is taken up or adsorbed onto other materials, this process may take some days. The three most commonly used fumigants were methyl bromide, phosphine and ethylene oxide. Other fumigants used in museums in the past in some countries were hydrogen cyanide, carbon disulphide, carbon tetrachloride and ethylene dichloride. Because they are so toxic, the use of fumigants is restricted in most countries by complex safety regulations.

Ethylene oxide was widely used but it is banned for museum use in most countries because it is carcinogenic. In addition, the gas is given off for a long period of time after objects have been fumigated.

Methyl bromide was the most widely used fumigant but it is now implicated as an ozone-depleting gas. There are now restrictions placed in the use of this fumigant and it will soon be phased out in most countries. Many museums which relied on regular treatment with methyl bromide are now having to look for suitable replacement methods.

Phosphine has been used for treatment of museum objects in some countries. Although it is an effective fumigant which does not leave residues, it is slow-acting at low temperatures. In addition it will cause corrosion of metalwork at higher RH levels.

Sulphuryl fluoride is one of those currently being evaluated as a replacement for methyl bromide. It has been successfully used against termites in the USA for some years and is now registered for use in some other countries. There are concerns that impurities in the commercial gas can cause corrosion of metallic surfaces and this needs to be carefully checked before it is used in museums or historic houses.

All successful fumigation relies on a knowledge of the properties and limitations of the fumigant gas which is being used. It is particularly important that the complex relationship between the concentration of the gas and the time of exposure is understood. ***For this and reasons of safety to staff, fumigation must only be carried out by people who have been trained in the techniques.***

Fumigation had a particular value for control of some pests (e.g. wood-boring beetles) where it was the only practical method of controlling hidden infestation and immediately preventing further damage. However, contrary to some beliefs, fumigation conferred no protection from further attack once the gas was aired off. The term 'preventive fumigation' was often used by museums which routinely fumigated all objects destined for storage. The expression is misleading because fumigation will only kill insects present in objects and although it prevents infestations developing from this nucleus, it will not protect any object from re-infestation.

The effects of fumigant gases and consequent residues on objects do give cause for concern. Extensive airing-off periods are needed in order to disperse traces of gas which have been adsorbed by the fumigated goods. Specially constructed vacuum chambers have been used to reduce airing-off times and also give quicker and more effective fumigation.

Most museums and historic houses can use alternative methods such as freezing or anoxia for treatment of objects and no longer need to rely on fumigation.

Damage to objects by insecticides

The application of any persistent pesticide directly to objects will leave residues of that chemical or its breakdown products. Such residues may be unacceptable and this will therefore rule out the use of anything but the non-persistent pyrethroids. However, it is unlikely that intelligent application of residual pyrethroid, organophosphorus or carbamate insecticides to the museum or house structure will cause detectable residues in displayed or stored objects.

Dichlorvos vapour will permeate the atmosphere around objects and may cause some deleterious effects, particularly where concentrations are locally higher than recommended (Dawson 1988). For example, new dichlorvos slow-release strips may produce drops of liquid dichlorvos which may then drip onto objects causing bleaching of dyes and corrosion of metals. The same effects can occur if strips are used in confined spaces at far higher concentrations than those recommended on the label. Corrosion of pins used for mounting entomological specimens can occur in over-dosed drawers.

Wood treatments and **moth proofing agents** are designed to be applied directly to the material needing treatment or protection and some insecticides and formulations are particularly effective. The recent development of water-based micro-emulsion formulations allows such protective treatments to be carried out with far less risk to objects (Pinniger *et al.* 1994). However, the need for protection and control must be balanced against the deliberate contamination of objects with chemicals which will produce long-term residues and may alter the chemistry or appearance of objects (Toshiko 1977; Koestler *et al.* 1993).

The deleterious effects of **fumigant gases** on artefacts and natural history specimens has already been mentioned. There are specific examples of corrosion of metallic objects by phosphine and decrease in tensile strength of silk and cellulose materials following ethylene oxide fumigation. Problems of taint and damage following fumigation with methyl bromide have almost always been attributed to the effect of liquid methyl bromide contacting objects. This has usually been the result of poor fumigation practice and failure to ensure that the methyl bromide has been properly vaporised. Many objects may have chemical residues resulting from past treatments (see below).

In cases where the evidence shows a need for the chemical treatment of objects, seek advice from a trained conservator and always use products approved for use. If a specialist pest control company is employed to carry out treatments, ensure that they are aware of the implications of their use of pesticides.

After treatment

Remember, once the treatment is completed, objects should be checked to ensure that all insects have been killed and care must be taken not to allow the material to become re-infested. Keep treated and pest-free objects separate from actively infested materials or storage areas. In cool and dry storage areas, objects which are particularly susceptible to insect attack can be left in clear polyethylene or heat-sealed bags to prevent them from being re-infested but only use plastic material which is acceptable to conservators.

Safety and pesticide use

There are many concerns over the safety of current pesticides and also residues from treatments carried out in the past. Many pesticides which are now banned, or no longer in use, are very persistent and insecticide deposits may still be present on objects treated many years ago. Animal and bird skins were often treated with arsenical soaps. This is one of the reasons why many historic animal specimens have survived undamaged whereas other, more recent, specimens may be destroyed by insects. Herbarium specimens and mounting sheets were treated with mercuric chloride and LPCP (lauryl pentachlorophenate). Insecticide dusts, such as DDT and Lindane, were widely used on natural history specimens and ethnographic materials. Many structural timbers were treated with gamma-HCH (Lindane). Residues of these chemicals are very long-lasting and documentation of use was either very poor or non-existent. It is therefore sensible to take precautions when handling older material, particularly if it shows signs of staining, crystals or white dust. Protective gloves and a dust mask should be sufficient protection in most cases but if the contamination is obvious and excessive then objects should be handled in a fume cupboard. Vapour insecticides, such as naphthalene and paradichlorobenzene, were used in very large quantities in storing natural history collections. Because everything absorbs the vapour, these chemicals persist for many years after they are no longer used. Washing of treated textiles may release unpleasant fumes of naphthalene which were not previously detected.

The effects of pesticides on objects have already been dealt with, but there is also the possibility of deleterious effect of pesticides on humans. This is a very emotive subject which deserves very careful consideration. There are regulations in force in many countries which stipulate that only approved pesticides may be used in the museum situation. When used they must only be applied in a specified way and appropriate safety precautions must be taken. Although these precautions should be on the label, they are not always present or available to the user. It is essential that any person who intends to apply pesticides should obtain a copy of these recommendations from the supplier before they apply the pesticide. The person who handles and dilutes pesticide concentrates is more at risk from contamination and should have access to any protective clothing necessary. As a general rule, it is also sensible to wear appropriate protective clothing when carrying out treatments and also to exclude any unnecessary personnel from the area during treatment. With most modern formulations, such as micro-emulsions or micro-encapsulated sprays, once a treatment has been applied, the hazard from deposits is generally low. There is always a need to take precautions to prevent inhalation of vapour and dust or excessive handling of treated objects. The solvent present in some formulations is more hazardous than the pesticide and water-based products are safer to use.

Some countries do not have comprehensive regulations relating to the use of pesticides in museums. Staff working in those places should endeavour to obtain as much information as possible from manufacturers and distributors about any pesticide which they are considering using. Labels, hazard sheets or information on conditions of use should be available if the product is registered in the UK, USA, Germany or Japan, and these should be obtained. Most of the insecticides now registered for use in museums are also approved for use in food processing and storage or domestic/home situations, and therefore usually present a very low hazard if used in the correct way. However, it is the duty of any pesticide user to safeguard themselves, colleagues, objects and the environment from unnecessary contamination or hazard. Where possible, use pest prevention and monitoring strategies which minimise pesticide use.

CHAPTER SIX

Rodents and birds

Adrian Meyer

The living organisms dealt with so far in this book do not have backbones. Animals which have a backbone are called vertebrates and this group includes the mammals, birds, reptiles, amphibians and fishes.

Rodents, the major mammal pest species, are able to colonise museums and historic houses and in so doing, cause serious problems or damage. A number of species of birds also cause problems when they live in, on, or near buildings.

Rodents and birds differ from invertebrates in many ways; for instance, they are unlikely to be as abundant, but they make up for this in their potential to cause damage by their:

- strength
- voracious appetites
- relative adaptability
- mobility
- ability to exploit their environment
- fouling of the environment.

Nevertheless, they also require an environment which provides the three main requisites of food, water and harbourage.

Rodents

Rodents as pests

Rodents are probably the most successful and abundant mammals on earth. With some 1800 species worldwide, they occupy almost every conceivable environment from harsh desert to the frozen tundra – there is no place where this cosmopolitan group can be ignored. The vast majority of these species hardly come into contact with man at all and when they do, would probably prefer not to. Some however, collectively known as the commensal rodents, are able to exploit the environment that man creates and live at his expense, invade his properties, eat his food and damage his possessions. Over the centuries, they have taken advantage of human transport and trade routes and they have spread to inhabit and infest all parts of the world.

There are three species with a worldwide distribution which are the most important:

1. Norway or brown rat (*Rattus norvegicus*)
2. Ship, roof or black rat (*Rattus rattus*)
3. House mouse (*Mus domesticus*)

There are other important commensal species with more localised distribution, but the basic approach to the biology and control of all rodents is the same.

Rodent lives

Rodents have more similarities to man than we can care to admit: their likes and dislikes are similar and frequently their abilities to exploit available resources leave man far behind. Like us, they want food and because they are omnivores, they are able to exploit almost any food that comes their way. They require water, although mice may also obtain water from the food they eat. They need shelter and prefer to avoid extremes of hot and cold; museums and historic houses therefore provide ideal homes for commensal rodents.

Once established in a building, the population of rodents can expand rapidly to exploit the available resources. In so doing, they damage anything in that environment in a non-selective way – paintings, wooden artefacts, leather, cloth, food, books, are all susceptible to attack. The potential of rodents to cause extreme damage and loss bears no relationship to their relatively small size. The fabric of the building, including the electrical wiring, alarm systems and drains are all susceptible to damage. In a number of cases, rodent-damaged wiring has led to fire.

Apart from the direct damage caused by rodent infestations, they foul their environment with their droppings and urine causing damage and deterioration of objects and the museum environment. Mice will urinate everywhere and as rodents can transmit diseases such as food poisoning, they must not be tolerated. The discovery of rodent infestation in food preparation areas has led to the closure of museum catering services by environmental health officers. House mice live about ten months and in this time will excrete about 0.5 litres of urine. Norway rats live for about a year and will excrete about 5 litres of urine in this time!

Appearance and behaviour

Rodents are distinguished from other mammals by the characteristic form of their teeth. They have only a single pair of front teeth (incisors) in both their upper and lower jaws and no teeth between these and the chewing teeth (molars) at the back (Fig. 35). It is their teeth which have contributed so much to the success of rodents. Their incisors are exceptionally strong and, more importantly, grow throughout the life of the animal so that they are able to gnaw even very hard objects made of lead, aluminium, etc. Indeed, the name rodent derives from the Latin word *rodere* – to gnaw.

The main visual characteristics and measurements of mice and the two rat species are shown in Figures 36, 37 and 38 and Table 4.

Breeding potential

One of the things which makes rodents such formidable pests is their ability to reproduce extremely rapidly. Table 5 summarises their reproductive characteristics and potential. Given ideal breeding conditions and unlimited food, water and harbourage, a single pair of Norway rats can give rise to about 40 direct off-spring within a year. In addition, the members of the first litter are in turn able to breed within two and a half months of birth so that it can be predicted that the population from this original single pair can be several hundred strong within a year. Other species such as the house mouse can breed even faster and this highlights the potential problem that exists if rodents are allowed to gain access to the susceptible museum environment. The risk of serious damage clearly increases with the rapid population rise.

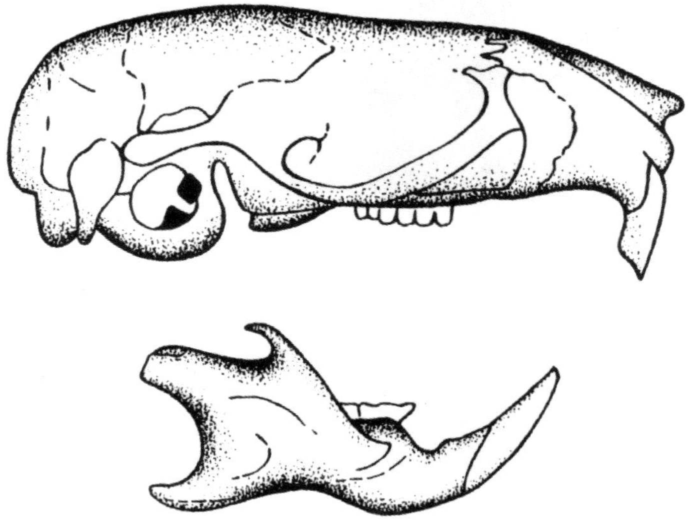

Figure 35 Rodent skull

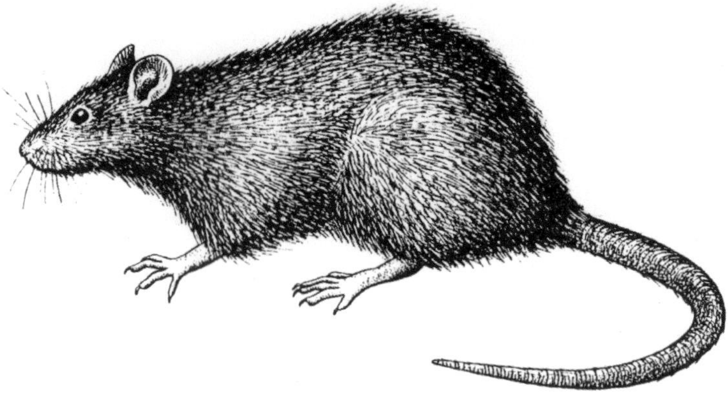

Figure 36 Brown rat

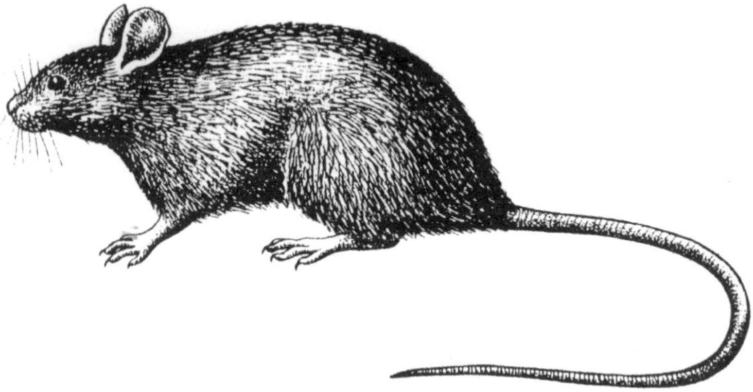

Figure 37 Black rat

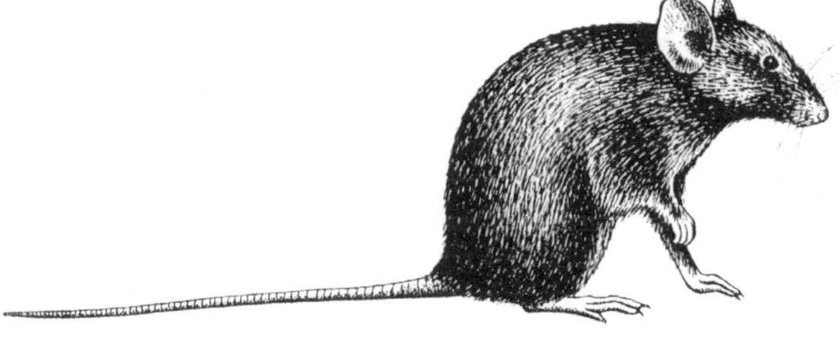

Figure 38 House mouse

Table 4 Characters and size of rodent pests

Character	Norway or brown rat *Rattus norvegicus*	Roof or black rat *Rattus rattus*	House mouse *Mus domesticus*
Weight	150–600gm	80–300gm	10–21gm
Head and body	Nose blunt, heavy, stocky body, 18–25cm	Nose pointed, slender body, 16–21cm	Nose pointed, slender body, 6–10cm
Tail	Shorter than head plus body, darker above and lighter below with short stiff hairs, 16–21cm	Longer than head plus body, uniformly dark coloured, naked, 19–25 cm	Equal to or a little longer than head plus body, uniformly dark coloured, naked, 7–11cm
Ears	Relatively small, close set, appear half-buried in fur. Rarely over 20–23mm	Large, prominent, thin and hairless, stand well out from fur, 25–28mm	Prominent, large for size of animal, 15mm or less
Fur	Brownish-grey on back, greyish on belly	Brownish-grey to black on back, belly may be white, grey, or greyish-black	Brownish-grey on back, greyish on belly
Droppings	Blunt-ended, 15–20mm long	Pointed and narrow, 10–12mm long	Pointed ends, 3–5mm long
Habits	Climbs, burrows, gnaws, usually lives outdoors, in sewers and drains, swims and dives	Agile climber, gnaws, often lives off the ground in trees etc., lives indoors and outdoors	Climbs, sometimes burrows, gnaws, lives indoors and outdoors

Table 5 Reproductive potential of female rodents

Species	*Rattus norvegicus*	*Rattus rattus*	*Mus domesticus*
Age at sexual maturity (days)	75	68	42
Gestation period (days)	22–24	20–22	19–21
Average number of young	9	6	6
Average number of litters per female/year	4/5	5/6	7/8
Age at weaning (days)	28	28	25
Average production per female/year	38	34	44

Senses

Rats and mice are mainly nocturnal animals and have developed skills to enable them to feed, explore, search for shelter and avoid predators.

- **Vision.** Rats and mice are able to recognise familiar objects in their environment and to orient themselves on these objects. Any unexpected movement, even in very dim light, will cause them to run rapidly to cover. Their safety depends upon familiarity with their surroundings and therefore they use the same part of their environment as much as possible.
- **Hearing.** Rodents have an acute sense of hearing and react to any sudden noise. They are able to detect sound and to communicate both in the

audible frequencies that man can hear as well as in the ultrasonic frequencies outside man's hearing.

- **Smell.** Rats and mice have a very sensitive sense of smell and are continually sampling the air in order to sense what is going on around them. They use smell markers to recognise both familiar and unfamiliar aspects of their environment as well as to recognise members of their own or other families.
- **Touch.** Their sense of touch is highly developed and their long sensitive whiskers are used to detect physical environmental changes and to help orient themselves within that environment.
- **Taste.** Commensal rodents are particularly sensitive to the food that they eat and to changes within the food. This sensitivity can lead to rejection of food and, depending upon the availability of alternative foods, has an important influence on their attitude to poison baits.

Other characteristics

- **Learning.** Rats and mice are very adaptive and can learn rapidly. This, together with their physical abilities, makes them adversaries to be respected.
- **Gnawing.** Rats and mice are able to gnaw through any material that is softer than the enamel of their teeth including most woods, aluminium sheeting, soft mortar and poorer quality asphalts and concrete. Galvanised sheet metals are not usually damaged and can therefore be used to proof buildings against rodent entry. Because of their exploratory behaviour, they gnaw things in addition to consumable food items and most materials in the museum environment are susceptible to attack.
- **Digging.** The Norway rat is the most prolific digger and burrows extensively when it is able to do so.
- **Climbing and jumping.** All commensal rodent species are able to climb, jump, swim and dive with varying degrees of success. Most rough walls can be easily scaled by rats, particularly the roof rat, but also by mice. Pipes, streams, sewers and walls are unlikely to present barriers to rodents.

Detection

The potential for serious damage occurs as soon as the first rodent gains access to the museum and increases as the population increases. The strategy must therefore be first to prevent the problem occurring and then to develop contingency plans for eliminating an infestation if one becomes established.

Effective prevention and control depends upon rodents being recognised and detected at as early a stage as possible. Because they are secretive and nocturnal, the animals themselves are not always seen, and their presence must be detected from the signs and traces that they leave behind.

Very often the first sign of an infestation is the discovery of droppings, damaged objects or half-eaten food and packaging. When this occurs, a survey should immediately be made of all areas where food might be available.

If the presence of a rodent infestation is confirmed, then a thorough survey should be undertaken to identify both the scope and size of the infestation and to identify the source. Clearly, an important part of any rodent control strategy is an awareness among the staff of the problems that rodents can cause and the signs and traces that indicate the presence of an infestation. Staff training and awareness should therefore form an essential part of any rodent survey strategy.

Surveying for rodents

All areas within the museum or house should be regularly inspected. Use a powerful flashlight and record all the positive findings. Those areas which are most likely to provide their basic requirements of food, water, and harbourage such as kitchens and food preparation areas, basements, drains, ducting, etc: should be treated as high risk areas and should be inspected frequently for signs of infestation.

- **Droppings** are a distinctive shape and size for each species (Fig. 39). They are prolific and may be found anywhere, but for rats in particular, they may be found in favoured areas such as in corners, on ledges and near their harbourage and food.

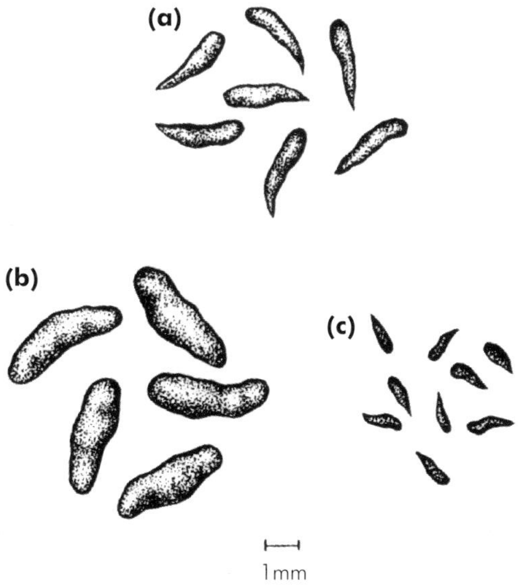

Figure 39 Rodent droppings: (a) Black rat; (b) Brown rat; (c) House mouse

- **Footprints and tail marks.** These can be found wherever rodents travel and there is dust. Search dusty places, under units, on shelves and ledges and outside in soil and mud as these are likely to be most productive.
- **Smears.** Rats and mice make use of the same routes in their wanderings and they leave behind dark coloured greasy smears on surfaces against which they pass. These are known as rodent smears: the stronger the smear, the heavier the infestation or the longer it has been present. They are most likely to be found on girders, around holes, on corners, and at the base of doors and windows.
- **Holes and damage.** All rodents gnaw or dig holes to provide access to harbourage or food. Any holes and damaged objects should be looked at closely to identify gnaw marks, footprints, half-eaten food or other signs of activity. Norway rats, in particular, dig readily in banks, flowerbeds, and under foundations or pavements.

Preventing rodent infestation

The objective of hygiene and proofing is to make the museum environment as inhospitable as possible for rodents.

Hygiene

Maintenance of good hygiene is vital if rodents are to be discouraged from living in, or moving into, museums and houses for shelter and food.

- Kitchens or areas where food is prepared or consumed must be cleaned daily and all waste or spilt food removed. All stored food should be kept in refrigerators, freezers or rodent-proof containers or cupboards. Special attention should be paid to cleaning inaccessible areas.
- Piles of rubbish should be cleared away from inside and outside the museum or house to remove potential harbourage. All storage areas should be kept tidy so that if rodents do find their way in, they do not encounter a hospitable environment and any rodent activity is immediately evident.
- Where it is aesthetically acceptable, undergrowth around the museum buildings should be cleared back for a distance of 3 metres from outside walls. Branches of trees should be cut if they are overhanging or in close proximity to the roof.
- High standards of cleanliness and tidiness should be maintained at all times.
- Additional areas that are 'high risk' and worthy of regular inspection include; basements, toilets, washing areas, attics and other areas with direct access to the outside.

Proofing

The proofing of any building – but particularly one as complex and varied as a museum or historic house – will present many problems. There may be particular difficulties because some of the most effective methods may be visually unacceptable in a historic building. It is important to remember that adult house mice can squeeze through holes of about 12mm and that young mice can get through gaps of 6mm.

- **External pipes.** Rodents may be prevented from entering and climbing the inside of rainwater and waste pipes by the use of back inlet gulleys. Where pipes discharge over a yard gulley, access to the eaves gutter may be prevented by fitting a wire balloon into the outlet from the gutter. Rodents can be prevented from climbing external downpipes by fixing flat or cone-shaped metal rat guards. Flat guards should have the edges turned down by 50mm.
- **Holes, ducts and internal pipes.** Any small holes in walls and floor should be filled with mortar. Large holes should be filled with brickwork or stone set in mortar or concrete. Holes in timber can be proofed by nailing sheet metal over the holes. Pipes and cables should be tightly fitted when passing through walls and floors. Metal flanges or plates should be fitted around the pipes and cables. If the internal diameter of the sleeve is such that space between it and the pipe is more than 6mm, it should be fitted with tightly packed, crumpled-up wire netting.
- **Sewers and drains.** Rats can live in large numbers in sewers and drains and infest the surface via breaks and faults in the drainage system. Consequently, every attempt should be made to ensure that such faults do not occur, however, in many cases they may be difficult to locate. If problems are suspected, specialist engineers should be called in. General maintenance work should involve the repair of inspection chambers and replacement of broken interceptor traps, caps to rodding arms, manhole covers and fresh air inlets. Defective WC pans should be replaced and water seals maintained properly. Dead lengths of drain should either be filled solid with concrete or the disused piping collapsed and the line of the drain back-filled.
- **Cavity walls.** These provide good harbourage for rats and mice. Access is often through air bricks and ventilators, and may be prevented by keeping openings to 6mm or less by covering them with metal mesh.
- **Doors.** These should fit tightly into the frame and need protection from gnawing at the base. If it is acceptable, this protection should be in the form of a metal kicking plate with similar plates fixed to the bottom of the frame or lining to form a continuous band of metal. Door thresholds should be level and must be built up if worn.

  Bristle strip attached to the bottom of doors can help prevent rodents getting in but it is not a complete barrier. The rodents can gnaw through the strip given the opportunity.

Rodent control

When rodents are found in a museum or house that should have been protected, then to some extent the rodent strategy can be said to have failed. Realistically, however, it is unlikely that prevention is always successful and the ability to eliminate the infestation through the use of poisons or other techniques is an essential secondary line of defence. If control measures only eliminate part of the infestation, then the mobility and reproductive potential of rodents will ensure that the problem rapidly returns.

The use of poisons is the only control method that is likely to prove regularly effective, although the use of traps and sticky boards can reduce the levels of infestation. The use of guns, snares, ultrasound, repellents and other techniques are not practical alternatives as they are not sufficiently effective to eliminate an infestation.

Traps

A wide range of types of traps is available. The most effective are likely to be the 'break back' types which are usually relatively simple to use. Their effectiveness will depend upon having a considerable number of traps set in the area where the rodents are moving and in having the traps positioned correctly. All traps should be placed with the food or treadle-end placed against a wall or solid barrier along which the rodents are moving. This will help to ensure that the rodents set the trap off as they walk over it. The inclusion of a food bait in traps may help, but is not essential (contrary to popular opinion, cheese is not the most attractive foodstuff to mice). Rodents that are hurt or frightened by the trap going off, but which escape, may become trap-shy and avoid traps in the future. Some individuals are trap-shy in the first place and are unlikely to be caught.

Live traps, both single and multi-catch, may be used in place of break back traps. Indeed, live capture traps may be more effective than break back traps because they are less likely to hurt or frighten the rodents, which do not then become trap-shy.

All traps should be inspected every day and any rodents caught must be removed and the traps reset. Trapping is most effective when used in combination with other control techniques.

Sticky boards

These work through the use of a sticky glue on boards which are placed in the area where the rodents are moving. Any rodents walking onto the glue are held and can be removed and humanely destroyed. This technique can be effective but is considered inhumane by some and should only be used if it is acceptable and codes of good practice are followed.

Poisons

All rodent poisons (rodenticides) are toxic, not only to rodents, but also to man and domestic and farm animals.

The use of rodenticides requires great care and should only be undertaken by trained staff or by professional pest control operators. Every precaution should be taken to ensure that the poisons are not made available to children or non-target animals and that clear warnings are given of any poison treatments to prevent anybody interfering with the baits.

There are two main types of rodenticides: acute and chronic.

Acute poisons are designed to kill the rodent after a single feed of the poisoned bait. The objective of the poisoning treatment is to encourage the rodents to eat as much of the poisoned bait in one feed as possible. Once eaten, the acute poisons have two characteristics which make them less than ideal for use as rodenticides:

- They take effect relatively rapidly (in hours rather than days).
- They cause painful and unpleasant symptoms as they take effect.

Because of these characteristics, any rodents which eat less than the dose required to kill them experience unpleasant symptoms before they recover. They will in future, show an aversion to the poison and the bait which is called bait-shyness.

Acute poisons will only rarely be used in museums or historic houses, the exception being the use of calciferol (a sub-acute rodenticide), which can be very effective when used against house mice.

Chronic poisons, unlike acute poisons, do not depend upon the rodents eating a single dose – they work most efficiently if taken in small doses over a number of feeds. In addition, unlike the acute poisons, they do not cause unpleasant symptoms and take some time to work. Therefore, if any rodents recover from a treatment that is inadequate, they do not show any shyness to future control operations and poisoning can be resumed or continued with a high expectation of success. For these reasons, when correctly used, chronic poisons can be expected to clear the infestation and provide a better level of control than acute poisons.

The most commonly used chronic poisons are anticoagulants which work by preventing blood clotting. They include: warfarin, chlorophacinone, diphacinone, pindone, difenacoum, bromadiolone, difethialone, flocoumafen and brodifacoum. All should be presented to the rodents in a suitable bait for a period of at least three weeks.

Chronic rodenticides can be applied in a number of ways including dusts and gels, but edible baits are the most widely used. Most of the anticoagulants are available in ready-to-use edible bait formulations and are probably most conveniently used in this form. Wax blocks may be more convenient and avoid the risk of spillage, but are not as palatable as loose-cereal baits. If loose-grain baits are used great care must be taken to prevent spillage as this can be a source of food for insects. Baits should be laid in containers or under cover to allow rodent access but to prevent access by non-target animals and children. Baits can be placed under furniture or cupboards, in lengths of drainpipe, under boards or in specially designed protected bait boxes.

The effectiveness of all the above rodenticides depends upon correct placement. The formulations should be placed throughout the area covered by the infestation at sites where they will intercept rodent movement. It is also essential to ensure that baits are placed at the correct density, perhaps every 2m for effective control of a heavy house mouse infestation.

Remember that rodents are three-dimensional animals and it is essential to place baits off the ground if there is evidence of vertical activity.

The length of time required to achieve control will vary according to the products used and the efficiency with which they have been laid. The chronic anticoagulants can be expected to achieve control within three to four weeks. Calciferol, when used for house mouse control, can be expected to achieve control in three weeks.

Resistance

This is the development of groups of animals which can tolerate higher doses of poison. When this ability is passed on to subsequent generations, increasing numbers of the rodent population will survive treatments. Resistance to rodenticides is worldwide and can cause severe problems in some areas. Successful use of anticoagulants, particularly for the house mouse (*Mus domesticus*) will depend therefore upon the resistance status of the infestation to be controlled. Local health authorities or agricultural departments are likely to be aware of the resistance status of rodents in the area. In urban areas rat resistance is less common, although resistance in house mice is widespread.

In some urban areas there has been poor control because house mice avoid all cereal-based bait. If this 'behavioural resistance' is encountered, alternative non-cereal baits will have to be used.

One of the problems with an extensive rodent baiting programme is that the rodents will often die in inaccessible places. This can not only lead to unpleasant smells, but the mummified remains provide a substantial source of food for many insect pests. It is much better to nip the problem in the bud by using a few traps when rodents are first detected than leave it until there is a substantial population needing widespread baiting with rodenticide.

Pest control contractors

In most museums the rodent control will be contracted out to specialist companies. It is essential that the contract is managed by a member of the museum staff who has sufficient knowledge of pest control to be able to manage the contract effectively.

Only trained and qualified technicians should be used by the contractor. The contract should specify the species covered and should identify the time scales within which control is expected to be achieved. It is essential that the contract includes a thorough monitoring and inspection programme.

Insect infestation of rodent baits

Old rodent baits can cause serious insect problems. Infestations of grain weevil, saw-toothed grain beetle, biscuit beetle, spider beetle and various moth species have been found in old loose-grain baits. The use of wax block baits seemed to reduce the risk but recently there have been increasing problems in museums and historic houses with the woolly bear larvae of carpet beetles (*Anthrenus* sp.) living in wax blocks. These insects were breeding in the baits and then spreading into textiles and natural history specimens. Some technicians do not recognise insect attack on baits; the main cause is old baits left behind by contractors who have lost the contract for servicing a building.

To eliminate this problem, follow good pest control practice:

- Check all rodent baits regularly for signs of insect damage.
- If there are larval skins or insect frass on a bait, replace it and take it off-site for disposal.
- Remove and destroy all old rodent baits, even if they belong to a previous contractor.

Rodent control strategy

- Recognise the risk points and areas in the museum or house.
- Proof where possible.
- Keep hygiene standards high.
- Ensure that caterers know their responsibilities.
- Train staff to recognise rodents and rodent signs.
- Ensure that all staff report rodents.
- Set up a system of regular inspections, particularly in high risk areas.
- Apply appropriate control measures as soon as infestation is identified.
- Ensure that any pest control contractor brought in understands the problem and is applying the most appropriate treatment.

Birds

A number of bird species are particularly suited to living in the urban environment in which many museums and museum stores are situated. Some of these birds can develop very high population levels and have a significant effect on the buildings and people in and around them. Three species in particular can cause problems:

- The feral pigeon (*Columba livia*)
- The house sparrow (*Passer domesticus*)
- The starling (*Sturnus vulgaris*)

Other bird species may cause problems in some rural areas or in other countries but the biological principles and the solutions applied to solve the problems remain the same.

Feral pigeons

Feral pigeons can breed throughout the year. Nests are built in or on buildings, under eaves or any other structures that provide support and protection. They are usually found on ledges or in hollows and in time and with repeated use they can become filled with pigeon droppings and debris. Up to four broods may be reared during a year with one or two chicks per brood.

Feral pigeons normally feed in flocks and become highly efficient at taking food when it is left unattended for short periods. They rely mainly on spillage around food preparation areas or on food (including bread, cake and seeds) given, spilt or discarded by staff and members of the public. The size of a feral pigeon population in an area is directly related to the amount of food available. Double the amount of food and the population doubles; halve the food supply and the population is halved.

Feral pigeons transmit diseases, including ornithosis (a form of psittacosis) and there are proven cases of humans catching the disease from pigeons. Feral pigeons can also carry the bacteria responsible for many incidents of food poisoning (Salmonellosis) and, although its incidence is low, this public health hazard should not be ignored.

Most damage caused by feral pigeons arises from their infestation of buildings where populations can build up to high levels. Fouling and contamination of structures and monuments can occur where the birds nest, rest or roost (Rees 2001). This is not only unsightly, but can produce foul smells and cause allergic reactions in some people. Their acidic droppings cause erosion of stonework and mortar. Accumulations of feral pigeon droppings can become infested with large numbers of mites and insects and collections may become infested from this source. The mobility of feral pigeons means that they are able to transmit these infestations from one building to another. Debris and dead birds can also be a source of flies, beetles, and other insects, as well as blocking rainwater pipes with subsequent extensive damage.

Sparrows

House sparrows may build nests in many places, including holes in walls and trees, in rainwater pipes, under eaves – in almost any place that provides shelter. Sites in and around buildings are preferred. They are largely dependent on man for both their food supplies and their nesting sites, and can occur in large numbers in and around museums and stores in urban areas. Their small size enables them to enter buildings through small apertures and, like mice, they are hard to exclude. Their nests and debris can cause blockages in rainwater discharge systems and air-conditioning units. The decline in numbers of house sparrows in the UK in recent years has reduced the risks.

Starlings

Starlings breed in holes in buildings or trees. The nest is untidy and lined with grass, feathers or moss. There may be two broods in a season. Starling problems in museums are caused almost exclusively by the habit of this species of night roosting in very large flocks. In towns, starlings roost in trees or on the ledges and facades of buildings. The noise from the birds is sometimes troublesome and their droppings foul and contaminate the building structure. A roost of several hundred thousand starlings is not uncommon and can cause extensive damage in a relatively short period of time.

Management of birds

As with the control of most rodent and insect pests that cause loss or damage, emphasis must be upon reducing or eliminating the population. With birds, successful damage prevention is more likely to be achieved through environmental manipulation and prevention of access than by attempts to directly reduce the population. This is because the birds are very mobile and can easily and rapidly move into areas from which other birds have been removed. The high population of pest species in urban areas provides a reservoir of birds that can replace those removed.

It is very important to remember that in many countries strict laws may exist to protect bird species and restrict the ways in which they may be controlled. The first step of any control programme is to check with local authorities that the methods that are proposed do not contravene any laws.

Prevention

- **Food.** One of the main reasons that feral pigeons and house sparrows are attracted to an area is availability of food. Where at all possible, the public should be instructed to refrain from feeding the birds. Reduction in food supply remains the single most effective method of reducing bird populations locally, particularly feral pigeons.

  Look very closely at the bird population that is causing a problem and identify where it is feeding. If the birds are feeding within the area of the museum or in an area away from the museum but where some influence might be exerted, the first step is to ensure that this food source is eliminated.

- **Proofing.** Birds may enter buildings either inadvertently or intentionally seeking shelter and warmth. They may then foul the interior of the building, set off alarm systems and generally cause a nuisance. To prevent bird entry to a building, all openings in excess of about 20mm should be closed, including gaps between doors and thresholds. Roofs should be kept in good repair and holes or missing tiles replaced. The eaves of most buildings can be sealed by building up the outside wall to the underside of the roof covering. Where corrugated sheeting is used, rolled up hexagonal wire mesh should be forced into the gaps, or alternatively all spaces should be filled with a fine concrete mix. Fibrous materials, such as glass fibre and most foam plastics, are unsuitable.

  Where possible, permanently open windows, ventilators and any other openings should be proofed with wire mesh or netting. In cases where large sliding doors are kept open, suspended net screens can be used successfully. Suspended clear plastic strips can also be used in some situations allowing safe passage and additionally retaining heat.
- **Netting.** Problems in the 'wells' of buildings can be solved by blocking bird entry by either netting the top of the well, or, where pigeons are the problem, stretching wires across the top of the well at about 30cm spacing to prevent the pigeons flying into the well. This second technique is particularly useful when netting might exclude light. If birds are entering the museum, watch them to find out their preferred entry routes so that only those areas where entry is confirmed need be proofed.

  Netting is a very effective technique for excluding birds from almost all areas where their access may cause problems. It can be very cost-effective but requires maintenance. The use of nets simply involves placing bird-proof nets across any opening that needs to be blocked. The nets need to be applied correctly and no gaps should be left between the nets and the frame.

  It is essential that the correct size mesh is used and the following mesh sizes are appropriate for the following sized birds.

Species	Mesh size
Feral pigeons	50mm
Starlings	28mm
House sparrows	19mm

Netting can also be used to protect statues, ornamental stonework and alcoves.

Curtains

A range of curtain techniques are available including plastic strips, metal chains and air curtains.

Repellents and deterrents

Birds can be prevented from nesting and roosting in sheltered and suitable positions on the outside of buildings in a variety of ways although some of these methods may not be acceptable on buildings of historical importance. Most are effective against feral pigeons, but are not usually effective against starlings and house sparrows.

- **Gels.** A plastic gel is applied to the horizontal surfaces, particularly along the leading edges of sills and ledges along which the birds are roosting. This provides an unstable surface on which the birds are reluctant to land. However, the technique has a limited life and can soon become unsightly. It is not recommended for general use.
- **Bird (sprung) wires.** This technique involves the use of wires suspended horizontally between spring anchors along the leading edges of sills and ledges, and is most effective in areas where the birds are less committed to a particular spot. Birds in some areas have learnt to overcome this technique and perch behind the wires.
- **Spikes and baffles.** The use of blunted spikes on a base strip of metal or plastic can be effective. It is essential that the spike density is sufficient and that the spikes are well secured to the sills. This technique is now the most widely used system and can be used in a variety of situations.
- **Distress calls.** Considerable success has been achieved in moving starling roosts through the use of starling distress call recordings. There is also a distress call for feral pigeons. It is essential that these calls are used with reinforcement, such as food reduction, if they are to be most effective. When these are played to the birds, it unsettles them, forcing them to move elsewhere. The method is not applicable for house sparrows which do not have distress calls.

Direct population management

The mobility of birds and their high populations in urban areas makes the use of population reduction techniques, such as poisons and trapping, less productive and rarely cost-effective.

In some circumstances, shooting may be appropriate as it can be used to remove specific birds that are causing a problem. This may be a bird caught in a room or gallery, or a bird roosting or nesting at a particular site. Low velocity weapons, preferably airguns, should only be used by experts who must familiarise themselves with local firearms legislation and liaise with the local police. This method is unacceptable if there is a risk of damaging valuable ornamentation or the fabric of a building.

Other vertebrate pests

The commensal rodents and urban birds are those vertebrate pests which are likely to cause the greatest concern. There are, however, other vertebrate species that will cause problems in museums and historic sites across the world, including squirrels, mongoose, monkeys, snakes and lizards. Non-urban birds will also cause problems as well as domesticated animals, such as cats, that have gone feral.

The methods to be employed in removing these unwanted guests will depend very much on the status of the species locally and nationally, and on the nature of the species itself. The first option that must be considered is simply to trap the offending animals and remove them elsewhere to be released. Subsequently, action must be taken in the form of proofing and hygiene improvements making it impossible for the animals to enter the building again. If the problem proves more difficult to solve, the use of poisons may be considered, but only if local and national legislation permits their use.

Bats

Bats may be found roosting in attics and roof areas of older museums and historic houses. There are many species, but the pipistrelle is the most common in churches and other buildings. While some people may find their presence disturbing, they are not likely to cause any serious damage, nor transmit diseases. Their alkaline urine can cause damage to surfaces but dried droppings are not such a hazard. This is because they are mainly the undigested chitin from insect bodies and will not normally provide food for other pests. They are easily distinguished from rodent droppings as they crumble into shiny fragments when squashed or broken. In most situations there is no need to take any remedial action unless the bats are causing damage or they are living in facilities also used by the public or staff.

Exclusion by proofing is the best technique, but this should only be undertaken following specialist advice relating to the ecology of the bats. In the UK and some other countries bats are protected and no action may be taken against bats without prior approval. Advice should be sought (in the UK) from English Nature or the Bat Conservation Trust.

CHAPTER SEVEN

Implementing IPM

Museum practice

Existing museum practices and management will have a big influence on the implementation of IPM. Many of these can be used or adapted to provide better pest prevention and control.
Major factors include:

- Resources
- Collection policy
- Museum objectives and function
- Interaction with other museum and stores
- Design and display
- Age of collection and buildings
- Special constraints – political or historical

Resources

The levels of staffing of museums and stores will determine the numbers of people who are available to care for the collection and also the degree of expertise which can be deployed. Managers, curators, custodians, conservators, cleaners and support staff all have important roles to play. Good communication and co-operation between these groups can achieve greater improvements in collection care than almost any other influence. Division of responsibilities and friction between areas of specialisations which have their own priorities can lead to serious problems. It is therefore essential to understand these demands and provide a balanced view to achieve the best strategies for pest control.

Shortage of funding can lead to severe problems of staffing, maintenance and materials. However, increased funding may not necessarily solve problems unless the money is spent wisely. For example, the value and convenience of expensive and complex storage racking can be negated by installing it so that it creates new dead spaces for pests. Similarly, air-conditioning may encourage pest problems, rather than remove them, unless it is installed, operated and maintained in an appropriate way. With the financial limitations operating in most countries, it is essential to target any available money on the most fruitful areas.

Maintenance of the museum is also tied to staff resources and funding and it is unfortunate that in times of shortage, maintenance is often the first area to be cut. Although museums and houses may be appear to be attractive and well maintained, non-public areas and stores are frequently neglected. This can rapidly lead to conditions which encourage development of moulds, insects and rodents and to increased pest problems. Savings on maintenance may therefore be achieved only at the cost of serious and escalating pest problems.

Collection policy, museum objectives and functions

The type of objects which a museum collects, its objectives and its function will all influence the development of pest problems. It is unfortunate that many museums may not realise that their activities may be contributing to their pest problems.

Some museums do not appear to have a clear policy for collection and acquisitions. But unless some constraints upon quality, relevance, importance or quantity are imposed, the size of the collection and attendant problems may soon outstrip resources and preclude the proper implementation of measures to prevent pest infestation. A clear and specific collection policy must be applied to enable the existing collection to be properly cared for. It is also important to have a clear policy on the disposal of objects. Those which are seriously damaged by insects may present a serious threat to other parts of the collection. If collections policy allows, destruction or disposal may be the most economic option to solve the problem. This may be appropriate for seriously damaged natural history specimens. However, living infestation must not simply be transferred to someone else. Any objects which are given, for example, to education sections, must be properly disinfested.

The needs of a large museum with a general and wide-ranging national reference collection may appear to be different from a small historic house, but the principles of insect pest prevention and control should be identical. It is the nature of the objects and their storage environment which will influence their susceptibility to pests.

Table 6 Guidelines for IPM strategy

Initial action	• Nominate someone to initiate the pest prevention and control strategy. • Obtain a plan of the building or collection or make sketches. • Carry out a preliminary survey to identify high risk areas and objects. • Include the outside of the building, galleries, stores and any other areas. • Plan a more detailed inspection and trapping schedule. • Check existing pest control contracts. • Examine the movement of objects in and out of the museum. • Draft an outline IPM strategy.
Operation of the IPM strategy	• Explain to key people the objectives of IPM. • Form a small team to aid communication and spread the IPM load. For example, include staff from conservation, collection management and buildings management. • Check cleaning regimes. • Install insect traps at strategic places throughout museum display, conservation and store areas. • Regularly inspect high-risk objects. • Record the results of inspections, trap catches and the condition of all objects and the building on a regular basis. • Respond to any change in condition or detection of pest presence in the building or on objects. • Ensure that any control measures are taken by a responsible person who has consulted the appropriate experts. • Ensure that good maintenance and hygiene practices are adopted. • Reduce the opportunities for pest access to the museum and stores. • Maintain environmental conditions which discourage pest development. • Liaise with pest control contractors. • Plan a quarantine strategy if one is not in place. • Maintain quarantine procedures for new, on loan, or returned materials. • Train staff in pest awareness and IPM. • Identify budgets which may contribute funding for IPM, e.g. training, buildings maintenance, collections care, storage furniture etc. • Complete the formal strategy for short-term and long-term IPM. • The success of the IPM strategy should be reviewed on an annual basis and amended where necessary.

Interaction with other museums and collections

The operation of a clear accessions and loans system in a museum must be established because of the high risks from pest introduction on incoming objects. The need for quarantine of suspect objects must be clearly understood and accepted. Some very vulnerable collections, such as herbaria, operate a policy of treating every batch of plants before it is allowed into collections areas. This strategy may be more effective in preventing the introduction of pests than inspection and incubation. Incoming packaging material may also harbour pests and this risk must be considered in the strategy.

The spread of some insect pests can be directly linked with movement of collections between museums. This is particularly true of natural history collections when entomologists or botanists may also unwittingly transfer pests with their working material. There can be a more positive side in that exchange of information between museums on advances in pest prevention and control strategies can often be best disseminated by example.

Age of collection and buildings and special constraints

It may be thought that the older the building and the objects therein, the greater the potential for problems. This may be true for museum storage buildings which have particular historical significance and alterations may not be possible for aesthetic or conservation reasons. However, some buildings and objects have been so extensively treated with preservatives and poisons in the past that they still remain lethal to pests. In addition, because of molecular degradation, the material of many older objects is far less nutritious than more recent material. The sugar and starch in wood degrades over the years so that old structural timber or wooden objects may not provide sufficient nutrients for woodborers. Newer, untreated wood used for construction or repairs may be far more vulnerable. Similarly, new, untreated wool textiles may be far more nutritious than old wool which has lost essential oils and fats.

While it is true that many modern museums and stores with full air-conditioning, proofing and specialist storage cabinets are free of pest problems, others have overlooked some of the basic principles of pest management, thereby providing more modern, but perfectly adequate, food and harbourage for insects.

Insect pest problems in tropical and sub-tropical climates

Because the major influences on pest development are temperature and RH, countries which have climates with high temperatures and high RH levels will have far more serious insect pest problems. Higher risks may also be linked to greater economic constraints and difficulties of obtaining advice and pest control materials. In most tropical countries a wide range of insect pests is indigenous to the country and can readily gain access to the museum whereas in temperate climates, many pests do not occur outside and are restricted to a few heated buildings. As already stated, insects reproduce and multiply much faster under warm, humid conditions and therefore in the tropics it is far more difficult to limit pest development by manipulation of the environment. Additionally, buildings and display and storage furniture made from traditional organic building materials may in themselves be more vulnerable to pest attack.

The principles of pest management outlined in this book should be applied where possible but they can be modified to suit local needs and conditions

(Sully *et al.* 2001). Synthetic pesticides may not be available and so local and ethnic control methods, such as wrapping objects in neem leaves, may be used to protect specimens. The use of solar heat for treating robust objects may also offer a more practical solution than expensive freezers. Adapting local building design may be far more suitable than trying to impose standard specifications from buildings designed for temperate countries. Examples include using natural ventilation in place of expensive and inappropriate air-conditioning, and storing objects in containers made from local pest-tolerant materials. There may also have to be some compromises so that the most serious pest risks are identified and tackled. It is pointless spending a great deal of time and effort in controlling a few clothes moths if the rest of the building and collection is being neglected and attacked by termites.

Training and communication

The most valuable resource of a museum is its staff and it follows that effective pest management makes the best use of available personnel. Successful pest prevention and control relies upon good communication between all those responsible for the collection. It is essential to combine the different expertise of curators, conservators, managers and support staff. All have an important role to play. For example, cleaning and security staff are in daily contact with collections and are often the first to notice signs of pests or pest damage.

As previously mentioned, one person must be responsible for co-ordinating pest prevention and control in a museum or group of museums to ensure that a sensible strategy is devised and implemented. Their role is also to ensure that all the staff and visitors are made aware of the strategy and how it affects them.

A basic core of information can be obtained from books as well as leaflets and handouts from various companies and government bodies. Specialist advice can be acquired from training courses, consultants and other museum and research centres with a special interest in museum problems. Conservators will need to be able to identify the main pests, recognise pest damage and advise on suitable control measures. Maintenance, cleaning and security staff must be made aware of the importance of reporting poor environments, building defects and possible insect activity. Good communication relies upon regular feedback between groups, and consistent and accurate record-keeping.

- An annual summary of pest incidence and attack should be produced and sent to relevant staff.
- Records should be kept of pests found on returned or newly acquired objects.
- Treatments of objects should be documented.
- Treatments carried out by contractors should be recorded.

- Continuity is essential and the loss of staff who have not kept records can lead to complete failure of the pest strategy and all the effort which has been previously expended.

Pest control contractors

When part of the pest prevention and control strategy involves the use of outside specialist pest control servicing companies, it is essential that communication is two-way. Most pest control contracts are dealt with by buildings and estates management and it essential to examine these contracts closely. Many pest control contracts are not specifically geared to a museum and the contractor may not understand the specialised needs of collection care. A clearly laid-out contract should be agreed with defined objectives. The museum should inform the contractor of their special needs or limitations and the contractor should ensure that they understand the priorities. There are many constraints imposed by working with historic objects or in a historic building and any pesticides used by the contractor must be acceptable to the museum. Many which may be technically approved may not be suitable for use. It is pointless to expect a pest control contractor to control rodents if building faults are not rectified and hygiene standards are poor. Old rodent baits encourage insect infestation which can then spread into the collections and so all bait positions must be recorded and checked regularly and old baits removed. Failure to agree a suitable programme for the contractor and related functions leads to control failure, poor value for money and may cause risk to objects. Many museums and houses waste money on ineffective and inappropriate pest control contracts – the money could be much more wisely spent. A well planned and co-ordinated pest control contract will provide both parties with an efficient and cost-effective result.

Health and safety

The health of staff and the safety of objects and the environment are of paramount importance. Each country has its own laws and guidelines which must be observed. Specific points relating to UK pest control are:

- All use of chemicals must be justified and recorded as required under Control of Substances Hazardous to Health (COSHH) Regulations.
- Under the Control of Pesticides Regulations 1986, certain insecticides are approved for use in museums by museum staff.
- Under the Control of Pesticides Regulations 1986, many chemicals are approved for use only by professional operators.

- Always check pesticide labels and follow the instructions for use of protective clothing and application.
- Some museums have old stocks of pesticides. Some may now be banned and must not be used. They should be disposed of correctly as stated in the Control of Pesticides Regulations 1986.
- The presence of bats may preclude the use of certain treatments in roof spaces. This is covered in the Wildlife and Countryside Act 1981.

IPM and the future

There are exciting new developments for the detection, prevention and control of pests and it is important that those responsible for developing IPM in museums do not work in isolation but share their experiences with others.

References

Ackery, P. R., Chambers, J. and Pinniger, D. B. (1999) 'Enhanced pest capture rates using pheromone-baited sticky traps in museums stores', *Studies in Conservation* **44**, 67–71.

Adams, R. G. (1978) 'The first British infestation of *Reesa vespulae*', *Entomologist's Gazette* **29**, 73–5.

Anon. (1999) 'Termites and UK buildings: biology, detection and diagnosis', *BRE Digest* **443**(1).

Armes, N. J. (1988) 'The seasonal activity of *Anthrenus sarnicus* and some other beetle pests in the museum environment', *Journal of Stored Product Research* **24**, 29–37.

Bergh, J-E. (1998) 'Current research on anoxic insect methods in Scandinavia', *Proceedings 3rd Nordic Symposium on insect pest control in museums*, Stockholm.

Berkouwer, M. (1994) 'Freezing to eradicate insect pests in textiles at Brodsworth Hall', *Conservator* **18**, 15–22.

Binker, G. (2001) 'Application of carbon dioxide for pest control of buildings and large objects', *Proceedings 2001 – a pest odyssey*, London [in press].

Brokerhof, A. W. (1999) 'Low-oxygen treatment and solarisation of Probota iconostasis: alternative pest control methods in the field'. *ICOM Committee for Conservation 12th Triennial Meeting*, Lyon.

Butcher-Youghans, S. and Anderson, G. E. (1990) 'A holistic approach to museum pest management', *History News* **45**(3).

Child, R. E. (1988) 'Fumigation in museums: A possible alternative', *Museums Journal*.

Child, R. E. (1994) 'The Thermo Lignum process for insect pest control', *Conservation News* 72(9).

Child, R. E. (1999) 'Insect pests in archives: detection, monitoring and control', *Journal of the Society of Archivists* **20**(2), 141–8.

Child, R. E. and Pinniger, D. B. (1994) 'Insect trapping in museums and historic houses', *IIC 15th International Conference, Preventative Conservation, practice, theory and research*, Ottawa, Canada.

Cox, P. D., Pinniger, D. B. and Mueller, D. (1996) 'Monitoring populations of the webbing clothes moth, *Tineola bisselliella*, using pheromone lures', *2nd International Conference on Insect Pests in the Urban Environment*, Edinburgh.

Crisafulli, S. (1980) 'Herbarium insect control with a freezer', *Brittonia* **32**, 224.

Daniels, V. (1997) 'The construction of a chamber for nitrogen fumigation', *The Conservator* **21**.

Dawson, J. (1988) 'Effects of insecticides on museum artifacts and materials', in L. A. Zycherman and J. R. Schrock, *A guide to museum pest control*, Washington, DC: American Institute for Conservation of Historic and Artistic Works and The Association of Systematics Collections.

Edwards, J. (1993) 'Biorationals and biotechnology – their future role in the control of insects in the urban environment', *International Conference on Insect Pests in the Urban Environment*, Cambridge, UK.

Edwards, R. and Mill, A. E. (1986) *Termites in buildings*. East Grinstead: Rentokil Library, Rentokil Ltd.

Elert, K. and Maekawa, S. (1997) 'Rentokil bubble in nitrogen anoxia treatment of museum pests', *Studies in Conservation* **42**, 247–52.

Finnegan, D. E. and Chambers, J. (1993) 'Identification of the sex pheromone of the Guernsey carpet beetle *Anthrenus sarnicus*', *Journal of Chemical Ecology* **19**, 971–82.

Florian, M-L. E. (1986) 'The Freezing Process – Effects on insects and artifact materials', *Leather Conservation News* 3(1), 1–13.

Florian, M-L. E. (1990) 'The effects of freezing and freeze-drying on natural history specimens', *Collection Forum* 6(2), 45–52.

Gilberg, M. (1989) 'Inert atmosphere fumigation of museum objects', *Studies in Conservation* **34**, 80–84.

Gilberg, M. (1990) 'Inert atmosphere disinfestation using Ageless^{R} oxygen scavenger', ICOM Committee for Conservation 9th Triennial Meeting, Dresden, GDR, 812–16.

Gilberg, M. (1991) 'The effects of low oxygen atmospheres on museum pests', *Studies in Conservation* **36**, 93–8.

Gilberg, M. and Brokerhof, A. (1991) 'The control of insects pests in museum collections: the effects of low temperatures on *Stegobium paniceum* the drugstore beetle', *Journal of the American Institute for Conservation* **30**, 197–201.

Gilberg, M. and Roach, A. (1991) 'The Use of a Commercial Pheromone Trap for Monitoring *Lasioderma serricorne* (F.) Infestations in Museum Collections', *Studies in Conservation* **36**, 243–7.

Hanlon, G., Daniel, V., Ravenel, N. and Maekawa, S. (1992) 'Dynamic system for nitrogen anoxia of large museum objects: a pest eradication study', *2nd International Conference on Biodeterioration of Cultural Property*, Yokohama, Japan.

Hansen, L. S. (1998) 'The common furniture beetle in Danish churches – pest densities and moisture conditions monitored for three years', *Proceedings 3rd Nordic Symposium on insect pest control in museums*, Stockholm.

Hansen, L. S. and Vagn Jensen, K-M. (1996) 'Upper lethal temperature limits of the common furniture beetle *Anobium punctatum*', *International Biodeterioration and Biodegradation*, Elsevier, 225–32.

Hansen, L. S. and Vagn Jensen, K-M. (1998) 'Heat treatments of the common furniture beetle at temperatures between 45°C and 54°C under dry and humid conditions', *Proceedings 3rd Nordic Symposium on insect pest control in museums*, Stockholm.

Hanus, J. (1985) 'Gamma radiation for use in archives and libraries', *Abbey Newsletter* 9(34).

Horokova, H. and Martinek, F. (1984) 'Disinfestation of archive documents by ionising radiation', *Restaurator* **6**.

Kigawa, R. *et al.* (2001) 'Practical methods of low oxygen atmosphere and carbon dioxide treatments for eradication of insect pests in Japan', *Proceedings 2001 – a pest odyssey*, London [in press].

Koestler, R. J., Parreira, E., Santoro, E. D. and Noble, P. (1993) 'Visual effects of selected biocides on easel painting materials', *Studies in Conservation* **38**, 265–73.

Koestler, R. J., Sardjono, S. and Koestler, D. L. (2000) 'Detection of insect infestation in museum objects by carbon dioxide measurements using FTIR', *International Biodeterioration and Biodegradation*, Elsevier.

Linnie, M. J. and Keatinge, M. J. (2000) 'Pest control in museums: toxicity of *para*-dichlorobenzene, Vapona and naphthalene against all stages in the life cycle of museum pests *Dermestes maculatus* and *Anthrenus verbasci*', *International Biodeterioration and Biodegradation* **45**, 1–13.

Miller, R. B. and Rajer, A. (1994) 'Freezing the herbarium: A novel approach to pest control', *IIC 15th International Conference, Preventative Conservation, practice, theory and research*, Ottawa, Canada.

Morgan, C. P., Pinniger, D. B. and Bowden, N. S. (1993) 'The effectiveness of residual insecticides against the varied carpet beetle *Anthrenus verbasci* and the implications for the control of this pest in museums', *International Conference on Insect Pests in the Urban Environment*, Cambridge, UK.

Nesheim, K. (1984) 'The Yale non-toxic method of eradicating book-eating insects by deep-freezing', *Restaurator* **6**, 147–64.

Newton, J. (1993) 'Carbon dioxide as a fumigant to replace methyl bromide in the control of insects and mites damaging stored products and artefacts', *International Conference on Insect Pests in the Urban Environment*, Cambridge, UK.

Newton, J., Abey-Koch, M. and Pinniger, D. B. (1996) 'Controlled atmosphere treatment of textile pests in antique curtains using nitrogen hypoxia – a case study', *2nd International Conference on Insect Pests in the Urban Environment*, Edinburgh, UK.

Peacock, E. R. (1979) '*Attagenus smirnovi*, a species new to Britain with keys to adults and larvae of British *Attagenus*', *Entomologist's Gazette* **30**, 131–6.

Petersson, K. (1998) 'Freezing as an insect control method on large clothes moth infestation', *Proceedings 3rd Nordic Symposium on insect pest control in museums*, Stockholm.

Philbrick, C. T. (1984) 'Comments on the use of the microwave as a method of herbarium insect control – possible drawbacks', *Taxon* **33**.

Pinniger, D. B. (1996) 'Insect control with the Thermo Lignum treatment', *Conservation News* **59**, 27–8.

Pinniger, D. B., Blyth, V. and Kingsley, H. (1998) 'Insect trapping: the key to pest management', *Proceedings 3rd Nordic Symposium on insect pest control in museums*, Stockholm.

Pinniger, D. B. and Child, R. E. (1996a) 'Woodworm – a necessary case for treatment? New techniques for the detection and control of furniture beetle', *2nd International Conference on Insect Pests in the Urban Environment*, Edinburgh, UK.

Pinniger, D. B. and Child, R. E. (1996b) 'Insecticides: optimising their performance and targeting their use in museums', *3rd International Conference on Biodeterioration of Cultural Property*, Bangkok, 1995.

Pinniger, D. B., Morgan, C., Child, R. E. and Lankford, W. (1994) 'A novel microemulsion formulation of permethrin for the control of museum insect pests', *IIC 15th International Conference, Preventative Conservation, practice, theory and research*,

Ottawa, Canada.
Quarles, W. (1997) 'Baiting subterranean termites', *IPM Practitioner* **XIX**(10).
Rees, S. (2001) 'Feral pigeons – a forgotten pest', *Proceedings 2001 – a pest odyssey,* London [in press].
Ridout, B. V. (2001) 'Understanding and controlling anobiid beetles', *Proceedings 2001 – a pest odyssey*, London [in press].
Rossol, M. and Jessup, W. C. (1996) 'No magic bullets: safe and ethical pest management strategies', *International Journal of Museum Management and Curatorship* **15**(2), 145–68.
Rust, M. K., Daniel, V., Druzik, J. R. and Preusser, F. D. (1996) 'The feasibility of using modified atmospheres to control insect pests in museums', *Restaurator* **17**(1), 43–60.
Selwitz, C. and Maekawa, S. (1998) *Inert gases in the control of museum insect pests.* Research in Conservation. The Getty Conservation Institute.
Shaw, M. R. (1999) '*Trogoderma angustum*, a museum and herbarium pest new to Britain', *Entomologist's Gazette* **50**, 99–102.
Stansfield, G. (1985) 'Pest control: A collection management problem', *Museums Journal* **85**(2), 97–9.
Strang, T. J. K. (1992) 'A review of published temperatures for the control of insect pests in museums', *Collection Forum* **8**(2), 41–67.
Strang, T. J. K. (1996) 'The effect of thermal methods of pest control on museum collections', *3rd International Conference on Biodeterioration of Cultural Property*, Bangkok, 1995.
Strang, T. J. K. (2001) 'Principles of heat disinfestation', *Proceedings 2001 – a pest odyssey*, London [in press].
Stusek, P., Pohleven, F. and Capl, D. (2000) 'Detection of wood boring insects by measurement of oxygen consumption', *International Biodeterioration and Biodegradation*, Elsevier.
Su, N-Y., Thomas, J. D. and Scheffrahn, R. H. (1998) 'Elimination of subterranean termite populations from the Statue of Liberty national monument using a bait matrix containing an insect growth regulator hexaflumuron', *Journal of the American Institute for Conservation* **37**.
Sully, D., Man-Yee, L. and Swee Mun, L. (2001) 'A topical solution to a tropical problem', *Proceedings 2001 – a pest odyssey*, London [in press].
Toshiko, K. (1977) 'Effects of insecticidal and fungicidal agents on materials of cultural properties', Scientific papers on Japanese Antiques and Art Crafts, No. 19–20.
Valentin, N. (1993) 'Comparative analysis of insect control by nitrogen, argon and carbon dioxide in museum, archive and herbarium collections', *International Biodeterioration and Biodegradation* **32**, 263–78.
White, P. R. and Birch, M. C. (1987) 'Female sex pheromone of the common furniture beetle *Anobium punctatum*', *Journal of Chemical Ecology* **13**(7), 1695–706.
Xavier-Rowe, A., Imison, D., Knight, B. and Pinniger, D. (2000) 'Using heat to kill museum insect pests – is it practical and safe?', *IIC Advances in Conservation*, Melbourne.

Further reading

Akerlund, M. (1991) *Angra – finns dom...?* Uppsala: Svenska museiforeningen. [In Swedish with good colour illustrations of insects]

Akerlund, M., Flato, S. and Hellekant, A. (1998) *Fran silverfisk till halsorisk.* Stockholm: LTs. [In Swedish with good colour illustrations of insects]

Allsopp, D. and Seal, K. J. (1986) *Introduction to biodeterioration.* London: Edward Arnold.

Bravery, A. F., Berry, R. W., Carey, J. K. and Cooper, D. E. (1987) *Recognising wood rot and insect damage in buildings.* Building Research Establishment Research Report. Garston: BRE.

Brokerhof, A. W. (1989) *Control of fungi and insects in collections of cultural value.* Amsterdam: Central Research Laboratory.

Buckle, A. P. and Smith, R. H. (eds) (1994) *Rodent Pests and their Control.* Wallingford: CAB International.

Busvine, J. R. (1986) *Insects and hygiene.* London: Chapman and Hall.

Carter, D. and Walker, A. K. (1999) *Care and conservation of natural history collections.* Oxford: Butterworth-Heinemann.

Cornwell, P. B. (1979) *Pest Control in Buildings.* East Grinstead: Rentokil Library, Rentokil Ltd.

Creffield, J. W. (1991) *Wood destroying insects, wood borers and termites.* Melbourne, Australia: CSIRO.

Croat, T. B. (1978) 'Survey of herbarium problems', *Taxon* 27, 203–18.

Daniel, V. and Hanlon, G. (1996) 'Non-toxic methods for pest control in museums', *3rd International Conference on Biodeterioration in Cultural Property*, Bangkok, 1995.

Dawson, J. E. (revised by J. K. Strang) (1992) *Solving Museum Insect Problems: Chemical control.* Canadian Conservation Institute Technical Bulletin 15. Ottawa: CCI.

Flieder, F. and Capderou, C. (1999) *Sauvegarde des collections du Patrimoine*, Paris: CNRS Editions. [In French]

Florian, M-L. E. (1987) 'Methodology used in insect pest surveys in museum buildings – a case history', ICOM Committee for Conservation, 8th Meeting, Sydney, Australia.

Florian, M-L. E. (1997) *Heritage Eaters – Insects and Fungi in Heritage Collections.* London: James and James.

Gallo, F. (1985) *Biological factors in the deterioration of paper.* Rome: ICCROM.

Goldberg, L. (1996) 'A history of pest control measures in the anthropology collections, National Museum of Natural History, Smithsonian Institution', *Journal of the American Institute for Conservation* **35**, 23–43.

Hall, A. R., Kenward, H. K. and Williams, D. (1980) 'Environmental evidence from Roman deposits in Skeldergate', *The Past Environment of York* **14**(3), 101–50.

Halstead, D. G. H. (1975) 'Changes in the status of insect pests in storage and domestic habitats', *Proceedings 1st International Conference on Stored Product Entomology*, Savannah, USA, 142–53.

Hickin, N. (1985) *Bookworms.* London: Sheppard Press.

Hillyer, L. and Blyth, V. (1992) 'Carpet beetle – a pilot study in detection and control', *Conservator* **16**, 65–77.

HMSO (2000) *Pesticides 2000.* Reference book. London: Ministry of Agriculture, Fisheries and Food/Health and Safety Executive.

Humptulip, W. (1839) *Howe to kille insects.* Ankh Morpork: Unseen University Press.

Linnie, M. J. (1987) 'Pest control: a survey of natural history museums in Great Britain and Ireland', *International Journal of Museum Management and Curatorship* **6**(3), 277–90.

Linnie, M. J. (1996) 'Integrated pest management: A proposed strategy for natural history museums', *International Journal of Museum Management and Curatorship* **15**(2), 133–43.

Mallis, A. (1997) *Handbook of pest control*, 8th edn. Cleveland: Frasak and Foster.

Meehan, A. P. (1984) *Rats and Mice. Their Biology and Control.* East Grinstead: Rentokil Ltd.

Meyer, A. N. (1998) 'Rodents in museums: not only a gnawing problem!', *Proceedings 3rd Nordic Symposium on insect pest control in museums*, Stockholm.

Mound, L. (1989) *Common insect pests of stored food products. A guide to their identification.* Ed. British Museum (Natural History) Economic Series No. 15. London: British Museum.

Mourier, H. and d'Aguilar, J. (1996) *Animaux et insectes hotes caches de nos maisons i Delachaux et Niestle, Laussanne.* Suisse. [In French with good colour illustrations of insects, birds and rodents]

Mourier, H. and Winding, O. (1986) *Collins guide to wildlife in house and home.* London: Collins.

Norton, R. E. (1996) 'A case history of managing outbreaks of webbing clothes moth *Tineola bisselliella*', ICOM Committee for Conservation 11th Triennial Meeting, Vol. 1 Preventive Conservation.

O'Connor, J. P. and Ashe, P. (2000) *Irish indoor insects.* Dublin: Town House.

Parker, T. A. (1988) *Studies on integrated pest management for libraries and archives.* Paris: UNESCO.

Peacock, E. R. (1993) 'Adults and larvae of hide, larder and carpet beetles and their relatives', *Handbooks for the Identification of British Insects* **5**(3). London: Royal Entomological Society.

Pearson, C. (1993) 'Building out pests', *AICCM Bulletin* **19**(1&2).

Pinniger, D. B. (1997) 'Recent advances in the control of museum insect pests based on detection and targeted treatment', *Pesticide Outlook* (April), 15–20.

Reagan, B. M. (1982) 'Eradication of insects from wool textiles Part I', *Journal of the American Institute for Conservation* **21**(2), 1–34.

Ridout, B. (2000) *Timber decay in buildings – the conservation approach to treatment*. London: E & F N Spon.

Story, K. O. (1986) Approaches to pest management in museums. Suitland, Maryland: Smithsonian Institution.

Strang, T. J. K. (1998) 'Another brick in the wall', *Proceedings 3rd Nordic Symposium on insect pest control in museums*, Stockholm.

Vagn Jensen, K-M. and Hansen, L. S. (1998) 'Evaluation of chemical methods for the prevention of damage to textiles due to Dermestidae and *Tineola bisselliella*, *Proceedings 3rd Nordic Symposium on insect pest control in museums*, Stockholm.

Walker, K. and Bacon, L. (1987) 'A condition survey of specimens in the Horniman Museum – a progress report', in *Recent advances in the conservation and analysis of artifacts, Proceedings of the Institute of Archaeology Jubilee Conservation Conference*, London.

Winks, R. G. and Champ, B. R. (1977) 'The principles of pest control in museums', Conservation in Australia.

Yasutomi, K. and Umeya, K. (1995) *Household pests*. Tokyo: Zenkoku Nosun Kyoiku Kyokai. [In Japanese with many good colour illustrations of insects]

Zycherman, L. A. and Schrock, J. R. (eds) (1988) *A guide to museum pest control*. Washington, DC: American Institute for Conservation of Historic and Artistic Works and The Association of Systematics Collections.